CAMBRIDGE TRACTS IN MATHEMATICS

General Editors

B. BOLLOBAS, P. SARNAK, C.T.C. WALL

114   Spinors in Hilbert Space

R.J. Plymen
University of Manchester

P.L. Robinson
University of Florida

Spinors in Hilbert Space

CAMBRIDGE
UNIVERSITY PRESS

Published by the Press Syndicate of the University of Cambridge
The Pitt Building, Trumpington Street, Cambridge CB2 1RP
40 West 20th Street, New York, NY 10011–4211, USA
10 Stamford Road, Oakleigh, Melbourne 3166, Australia

First published 1994

Printed in Great Britain at the University Press, Cambridge

Library of Congress cataloguing in publication data available
British Library cataloguing in publication data

Plymen, Roger J.
  Spinors in Hilbert space / Roger Plymen, Paul Robinson.
    p.   cm. – (Cambridge tracts in mathematics: 114)
  Includes bibliographical references and index.
  ISBN 0 521 45022 5
  1. Spinor analysis. 2. Hilbert space. I. Robinson, P. L. (Paul
Lee), 1958– . II. Title. III. Series.
  QA433.P63 1994
  512' .57--dc20 94–27645CIP

ISBN 0 521 45022 5 hardback

To Hilary and Tohien

CONTENTS

PREFACE

In this tract we set forth an account of the various Clifford algebras attached to a real Hilbert space, incorporating a detailed study of their Fock representations and the relationships between them. Rather than attempt to be encyclopaedic, we concentrate upon topics that appear now to be established in essentially definitive form, unlikely to simplify to any appreciable degree. In fact, we take this opportunity to offer simplified proofs of many standard results in the theory. We presuppose an understanding of basic functional analysis: the only moderately exotic prerequisite is a familiarity with the fundamentals of operator algebras; however, we provide an appendix assembling these fundamentals for convenience. Given these prerequisites, along with the usual mathematical maturity, the body of the text is entirely self-contained. As a result, the book should be accessible to graduate students: indeed, the material is carefully presented with them very much in mind. To complement the body of the text, each chapter ends with a section of Remarks. These address related topics whose development usually calls for substantial detours from the main text; this being so, they come with references rather than with proofs. They also include historical commentary, intended as a guide to further study and certainly not meant to be comprehensive. Of course, we hope that this book will interest not only mathematicians but also theoretical physicists: after all, Fock representations underlie the standard model of a free fermion field.

The term spinor is traditionally applied to a vector in the space on which a spin representation acts. Over an even-dimensional real Hilbert space, all irreducible representations of the complex Clifford algebra are equivalent and all are referred to as spin representations. In infinite dimensions, we shall view Fock representations as spin representations. Vectors in Fock space then become spinors in Hilbert space, hence the title of our book.

Our title is intentionally reminiscent of the remarkable book *Spinors in Hilbert Space* by P.A.M. Dirac (Plenum Press, 1974). By turns inspired and idiosyncratic, this book by Dirac is essentially an outline, upon which our book may be regarded as a rigorous detailed elaboration.

INTRODUCTION

In this book, we construct and study associative complex algebras that arise naturally from a real inner product space; in addition, we investigate some of their representation theory.

More specifically, we take a real vector space V with inner product $(\cdot \mid \cdot)$ and consider a variety of complex Clifford algebras to which it naturally gives rise. The most fundamental of these is purely algebraic: the plain complex Clifford algebra $C(V)$; briefly, this is the smallest unital associative complex algebra in which V is linearly embedded so that $v^2 = (v \mid v)\mathbf{1}$ whenever $v \in V$. Beyond this, we also consider a pair of operator algebras: the C^* Clifford algebra $C[V]$ and the vN Clifford algebra $\mathcal{A}[V]$; the first of these is essentially the enveloping C^* algebra of $C(V)$ for a canonical involution whilst the second is essentially the von Neumann algebra generated by $C(V)$ in the left regular representation. In each case, the Clifford algebra encodes the geometry of V: thus vectors in V are perpendicular if and only if they anticommute within the Clifford algebra.

In Chapter One we address some of the many structural questions that naturally present themselves in relation to these Clifford algebras. Referring specifically to the C^* Clifford algebra for convenience, it turns out that $C[V]$ is essentially insensitive to whether or not V is complete; thus it often does no harm to suppose that V is a real Hilbert space. It also turns out that $C[V]$ has scalar centre and is simple unless V is odd-dimensional, in which case $C[V]$ has two-dimensional centre and is the sum of two simple ideals. The C^* Clifford algebra $C[V]$ is naturally graded: the orthogonal transformation $-I$ on V extends to a unique (grading) automorphism γ of $C[V]$ whose fixed subalgebra $C^+[V]$ is the even C^* Clifford algebra. In fact, $C^+[V]$ is itself central and simple unless V is even-dimensional; indeed, if V is infinite-dimensional then the C^* algebras $C^+[V]$ and $C[V]$ are actually isomorphic. Not only minus the identity, but any orthogonal transformation g of V extends to an automorphism θ_g of $C[V]$; we refer to θ_g as a Bogoliubov automorphism. Now, it is natural to ask for conditions on g that are necessary and sufficient in order for the automorphism θ_g to be inner. Although this question relates to purely internal structure of the C^* Clifford algebra, it appears to be handled most readily by the external means of representation theory. In this first chapter we merely note that the grading automorphism γ is not inner unless V is even-dimensional, returning to consider the general Bogoliubov automorphism in the last chapter.

The heart of the book concerns a very special class of representations of the C^* Clifford algebra $C[V]$ when the real Hilbert space V is other than odd-dimensional: these might be referred to as spin representations (as indeed they are, when V is even-dimensional) but we shall refer to them as Fock representations. In Chapter Two we focus our attention on a single Fock representation. For its definition, this requires that V be provided with a unitary structure J: an orthogonal complex structure making V into a complex Hilbert space. Abstractly, the Fock representation π_J of $C[V]$ is (up to unitary equivalence) the unique star-representation π of $C[V]$ on a complex Hilbert space $\mathbb{H}$ containing a cyclic unit vector Ω annihilated by $\pi(v + iJv)$ whenever $v \in V$; we refer to $\mathbb{H} = \mathbb{H}_J$ as Fock space and to $\Omega = \Omega_J$ as Fock vacuum. In practice, we construct the Fock representation in terms of creators and annihilators satisfying the canonical anticommutation relations. It turns out that the Fock representation of $C[V]$ is irreducible; when restricted to $C^+[V]$ it decomposes as the sum of two inequivalent irreducibles, the half-spin representations. In Chapter Three we consider relationships between Fock representations. Thus, when J and K are unitary structures on V the Fock representations π_J and π_K of $C[V]$ are unitarily equivalent if and only if the difference $K - J$ is Hilbert-Schmidt. It follows that if V is infinite-dimensional then $C[V]$ has an uncountable infinity of inequivalent Fock representations; in marked contrast, if the dimension of V is even then all irreducible representations of $C[V]$ are equivalent. Also, if J is a unitary structure on V and if g is an orthogonal transformation of V then there exists a unitary operator U on $\mathbb{H}_J$ satisfying $\pi_J(\theta_g a) = U\pi_J(a)U^*$ for all a in $C[V]$ precisely when the commutator $gJ - Jg$ is Hilbert-Schmidt. The group $O_J(V)$ comprising all orthogonal transformations of V having Hilbert-Schmidt commutator with J is called the restricted orthogonal group, also denoted $O_{\mathrm{res}}(V)$. Needless to say, the Fock representations by no means account for all representations of the C^* Clifford algebra; however, they are without doubt the most important in many respects.

In Chapter Four we return to settle a problem raised in Chapter One: namely, that of deciding precisely which Bogoliubov automorphisms of each Clifford algebra are inner. For the plain complex Clifford algebra $C(V)$ we employ purely internal methods. For the C^* Clifford algebra $C[V]$ we employ Fock representations of $C[V]$ itself; for the vN Clifford algebra $\mathcal{A}[V]$ we employ one very special Fock representation of $C[V \oplus V]$. To each Clifford algebra we associate an ideal $\mathcal{I}(V)$ of bounded linear operators on V with the property that if g is an orthogonal transformation of V then the Bogoliubov automorphism θ_g is inner under precisely the following conditions: either $g - I \in \mathcal{I}(V)$ and $\ker(g + I)$ is even-dimensional; or $g + I \in \mathcal{I}(V)$ and $\ker(g - I)$ is odd-dimensional. For the plain complex Clifford algebra, the ideal $\mathcal{I}(V)$ comprises precisely all finite-rank operators, as might be expected on account of its purely algebraic nature. For $C[V]$ the ideal $\mathcal{I}(V)$ comprises precisely all trace-class operators; for $\mathcal{A}[V]$ the ideal $\mathcal{I}(V)$ comprises precisely all Hilbert-Schmidt operators. In each case, the group of orthogonal transformations for which the corresponding Bogoliubov automorphisms are inner acquires a double covering group constituted from unitary implementing elements. These double covering groups are models of pin groups and furnish the spin groups that give their name to this chapter. In all honesty, we do not delve into the structure of these spin groups: we only go so far as to construct them.

It will be seen that the dimension of V exerts considerable influence over the various associated Clifford algebras: in fact, there is a fundamental trichotomy according to whether the dimension of V is infinite, even or odd. The case of odd-dimensional V does not feature largely in our account: we are concerned primarily with cases in which the dimension of V is either even or infinite. Thus, if V is even-dimensional then its complex Clifford algebra $C(V)$ has up to equivalence exactly one irreducible representation; an automorphism of $C(V)$ is inner if and only if it is implemented in one (any) irreducible representation. When V is infinite-dimensional, the picture changes dramatically: if g is an orthogonal transformation of V then the Bogoliubov automorphism θ_g of $C[V]$ being inner amounts to a trace-class condition, whereas θ_g being implemented in Fock representations amounts to a (significantly weaker) Hilbert-Schmidt condition; the former circumstance relates to the (internal) construction of spin groups, whereas the latter pertains to (external) current groups. One remark on our approach to infinite dimensions is in order. For a variety of reasons, it is customary to assume separability of an infinite-dimensional real Hilbert space; we have been careful to offer arguments that are valid for all infinite dimensions, avoiding assumptions of separability.

1

CLIFFORD ALGEBRAS

In this opening chapter, we collect together fundamental properties of a variety of (complex) Clifford algebras attached to the real inner product space V. The plain complex Clifford algebra $C(V)$ is the universal unital associative complex algebra containing V as a real subspace with the property that if $v \in V$ then $v^2 = \|v\|^2 \mathbf{1}$; this algebra has a unique involution such that if $v \in V$ then $v^* = v$. The involutive algebra $C(V)$ carries a unique norm with the C^* property, that if $a \in C(V)$ then $\|a^*a\| = \|a\|^2$; the completion of $C(V)$ relative to this norm is a C^* algebra called the C^* Clifford algebra $C[V]$. The C^* algebra $C[V]$ has a unique (even) state τ with the (central) property that if $a, b \in C[V]$ then $\tau(ba) = \tau(ab)$; the von Neumann algebra generated in the corresponding cyclic representation of $C[V]$ is the vN Clifford algebra $\mathcal{A}[V]$. When V is finite-dimensional, these algebras coincide; when V is infinite-dimensional they are all different.

In §1 we present a detailed account of the plain complex Clifford algebra. We begin by studying the most immediate properties of $C(V)$ when V is arbitrary. We next consider $C(V)$ first when V is finite-dimensional and then when V is more particularly even-dimensional. After this, we approach $C(V)$ when V is infinite-dimensional by means of approximations via subspaces of V having finite (often even) dimension. Lastly, we comment on the structure of $C(V)$ when the dimension of V is odd. All of this material is quite standard and may be extracted from a number of sources; we include it here for completeness and as an introduction.

In §2 we develop the basic structure of the C^* Clifford algebra $C[V]$ when V is infinite-dimensional. The approach we adopt is only one of a number that are possible; brief references to some alternative approaches are given in the Remarks at the end of this chapter. In keeping with the aim expressed in the Introduction, we develop the fundamentals without the assumption that V be separable; again we refer to the Remarks for a few special comments in case V is separable.

In §3 we study the vN Clifford algebra $\mathcal{A}[V]$ when V is infinite- dimensional. As a matter of detail, we introduce it as the von Neumann algebra that arises by closing (in either operator topology, weak or strong) the range of the left regular representation of $C(V)$ on its Hilbert space completion relative to the inner product determined by its unique central state. Once again, we avoid the assumption that V be separable; if V is separable then $\mathcal{A}[V]$ is a version of the hyperfinite II_1 factor, for more on which see the Remarks.

1.1 Clifford algebras

Our primary aim in this opening section is to develop some of the purely algebraic structure of the complex Clifford algebra over a real inner product space. Accordingly, we make no completeness assumptions on the underlying real inner product space, which we allow to have arbitrary dimension.

Thus, let V be an arbitrary real vector space upon which $(\cdot \mid \cdot)$ is a positive-definite inner product and denote by $\| \cdot \|$ the corresponding norm. By a *Clifford map* on V we shall mean a real-linear map $f : V \to B$ into a unital associative complex algebra B such that if $v \in V$ then $f(v)^2 = \|v\|^2 \mathbf{1}$. In these terms, we define a *complex Clifford algebra* over V to be a unital associative complex algebra A together with a Clifford map $\phi : V \to A$ satisfying the following *universal mapping property*: that if $f : V \to B$ is any Clifford map, then there exists a unique algebra map $F : A \to B$ such that $F \circ \phi = f$.

As we now proceed to show, V always carries a complex Clifford algebra and any two complex Clifford algebras over V are naturally isomorphic.

Existence

We dispose of the existence problem for complex Clifford algebras by means of a standard tensor product construction. Denote by $V^{\mathbb{C}} = \mathbb{C} \otimes V$ the complexification of V: thus, $V^{\mathbb{C}}$ is obtained from V by extending from real to complex scalars. Let $T(V)$ stand for the full tensor algebra over $V^{\mathbb{C}}$: thus,

$$T(V) = \bigoplus_{r=0}^{\infty} T^r(V)$$

where $T^0(V) = \mathbb{C}$ and where if $r > 0$ then

$$T^r(V) = \overleftarrow{V^{\mathbb{C}} \otimes} \cdots \overrightarrow{\otimes V^{\mathbb{C}}}$$

is the r-fold complex tensor power of $V^{\mathbb{C}}$. Of course, $T(V)$ is a unital associative complex algebra with $\mathbf{1} = 1 \in \mathbb{C} = T^0(V)$ as multiplicative identity. Let $I(V)$ be the bilateral ideal of $T(V)$ generated by the subset

$$\{v \otimes v - (v \mid v)\mathbf{1} : v \in V \subset V^{\mathbb{C}}\}.$$

Finally, let A be the quotient algebra $T(V)/I(V)$ and let $\phi : V \to A$ be the map sending $v \in V \subset V^{\mathbb{C}} = T^1(V)$ to its coset modulo $I(V)$. It is plain both that A is a unital associative complex algebra and that $\phi : V \to A$ is a Clifford map. Now, let $f : V \to B$ be a Clifford map and extend to $f : V^{\mathbb{C}} \to B$ by complex linearity. The universal mapping property for the tensor algebra guarantees that f extends uniquely to an algebra map $T(f) : T(V) \to B$. The assumption that f is a Clifford map ensures that $T(f)$ vanishes on the ideal $I(V)$. Consequently, there exists a unique algebra map $F : A \to B$ such that

$$
\begin{array}{ccc}
T(V) & & \\
& \searrow^{T(f)} & \\
\Big\downarrow & & B \\
& \nearrow_{F} & \\
A & &
\end{array}
$$

is a commutative diagram, in which the vertical is the canonical quotient map. It is now evident that F is the unique algebra map from A to B satisfying $F \circ \phi = f$.

Uniqueness

That any two complex Clifford algebras over V are naturally isomorphic follows as usual from the universal mapping property. In fact, let A and A' be complex Clifford algebras over V with Clifford maps $\phi : V \to A$ and $\phi' : V \to A'$. Since $\phi : V \to A$ satisfies the universal mapping property and $\phi' : V \to A'$ is a Clifford map, there is a unique algebra map $\Phi' : A \to A'$ such that $\Phi' \circ \phi = \phi'$. Similarly, there is a unique algebra map $\Phi : A' \to A$ such that $\Phi \circ \phi' = \phi$. Now $\Phi \circ \Phi' : A \to A$ is an algebra map such that

$$\Phi \circ \Phi' \circ \phi = \Phi \circ \phi' = \phi$$

whence $\Phi \circ \Phi'$ is the identity map, on account of the universal mapping property for $\phi : V \to A$ applied to $\phi : V \to A$ itself. Similarly, $\Phi' \circ \Phi : A' \to A'$ is also the identity map. Consequently, Φ and Φ' are mutually inverse algebra isomorphisms.

Having thus established that the real inner product space V carries an essentially unique complex Clifford algebra, we may fix one and with impunity refer to it as the *complex Clifford algebra $C(V)$ of V*. Notice that the *Clifford property*

$$v \in V \quad \Rightarrow \quad \phi(v)^2 = \|v\|^2 \mathbf{1}$$

satisfied by the Clifford map $\phi : V \to C(V)$ implies that ϕ is necessarily injective. This being so, we shall feel free to suppress ϕ and to identify V with its image in $C(V)$ whenever convenient.

Theorem 1.1.1   *The complex Clifford algebra $C(V)$ is generated by its real subspace V satisfying the Clifford relations*

$$x, y \in V \quad \Rightarrow \quad xy + yx = 2(x \mid y)\mathbf{1}.$$

Proof   The tensor algebra $T(V)$ is of course generated by its real subspace $V \subset V^{\mathbb{C}} = T^1(V)$; as a result, the quotient algebra $C(V) = T(V)/I(V)$ is generated by its own copy of V. The Clifford property of $C(V)$ asserts that $v^2 = \|v^2\|\mathbf{1}$ whenever $v \in V$; the Clifford relations follow at once upon polarization, replacing v by $x + y$ when $x, y \in V$. $\square$

The *Clifford relations* just established have as a particular consequence the following fact: that if $x, y \in V$ then

$$(x \mid y) = 0 \quad \Leftrightarrow \quad xy + yx = 0$$

so that vectors in V are orthogonal if and only if they anticommute as elements of $C(V)$. This is but one manifestation of a theme that will be repeated throughout the course of our study: namely, that geometry in V is reflected by algebra in $C(V)$.

Now the universal mapping property for the complex Clifford algebra has certain standard functorial consequences. Fundamental among these is the fact that isometries between real inner product spaces give rise to homomorphisms between their complex Clifford algebras. Here, if $(\cdot \mid \cdot)$ and $(\cdot \mid \cdot)'$ are inner products on the real vector spaces V and V' then the linear map $g : V \to V'$ is said to be *isometric* in case $(gx \mid gy)' = (x \mid y)$ whenever $x, y \in V$. For the sake of clarity, let us reinstate the canonical embeddings $\phi : V \to C(V)$ and $\phi' : V' \to C(V')$ of the real inner product spaces in their complex Clifford algebras.

Theorem 1.1.2 *If $g : V \to V'$ is an isometric linear map then there exists a unique algebra map $\theta_g : C(V) \to C(V')$ such that*

$$\theta_g \circ \phi = \phi' \circ g.$$

Proof By virtue of its isometric nature, g when followed by the canonical embedding $\phi' : V' \to C(V')$ yields a Clifford map $\phi' \circ g : V \to C(V')$. The universal mapping property for $C(V)$ now provides a unique algebra map $G : C(V) \to C(V')$ with the property that $G \circ \phi = \phi' \circ g$. All that remains is to set θ_g equal to G. $\qquad\square$

When we once again suppress the canonical embeddings, this result may be formulated as saying that the linear isometry $g : V \to V'$ extends uniquely to an algebra map $\theta_g : C(V) \to C(V')$.

As usual, we shall let $O(V)$ signify the *orthogonal group* of V: thus, $O(V)$ comprises all isometric real-linear automorphisms of V. As a particular instance of the functorial property in the preceding theorem, each orthogonal transformation $g \in O(V)$ extends uniquely to define an automorphism θ_g of the complex Clifford algebra $C(V)$. We shall follow the custom of referring to θ_g as the *Bogoliubov automorphism* of $C(V)$ induced by g. If also $h \in O(V)$ then each of θ_{gh} and $\theta_g \circ \theta_h$ is an automorphism of $C(V)$ extending gh; it follows that $\theta_{gh} = \theta_g \circ \theta_h$. Thus, we in fact have a group homomorphism

$$\theta : O(V) \to \operatorname{Aut} C(V)$$

representing the orthogonal group by automorphisms of the complex Clifford algebra. This automorphic group representation and its descendants will feature quite prominently in what follows.

One particular Bogoliubov automorphism is of special importance and deserves a separate symbol: we denote by γ the Bogoliubov automorphism θ_{-I} induced by minus the identity; thus γ is the unique automorphism of $C(V)$ sending each element of V to its negative. Since the orthogonal transformation $-I$ has period 2, so also does the automorphism γ; accordingly, we refer to γ as the *grading automorphism* of $C(V)$. The subalgebra $\ker(\gamma - I)$ of $C(V)$ fixed pointwise by γ is called the *even complex Clifford algebra* $C^+(V)$ of V; the complementary subspace $\ker(\gamma + I) \subset C(V)$ on which γ acts as minus the identity is denoted by $C^-(V)$. In keeping with our referring to γ as the grading automorphism, we refer to elements of $C^+(V)$ as being *even* and to elements of $C^-(V)$ as being *odd*.

In addition to its grading, the complex Clifford algebra has a canonical antiautomorphism and a canonical conjugation, their product being a

canonical adjoint operation on the complex Clifford algebra. We take
each of these in turn.

Let us denote by $C(V)^0$ the algebra opposite to $C(V)$: thus, $C(V)^0$ is
$C(V)$ as a set, with precisely the same linear structure but with reversed
product, so that the identity map $C(V) \to C(V)^0$ is an antiisomorphism
of algebras. It is plain that the canonical inclusion $V \to C(V)^0$ is a Clifford map, this being the suppressed $\phi : V \to C(V)$ followed by the identity
map $C(V) \to C(V)^0$. The universal mapping property for $C(V)$ provides
a unique algebra homomorphism α from $C(V)$ to $C(V)^0$ restricting to
V as the identity; of course, we may view α as an antihomomorphism
from the algebra $C(V)$ to itself. The composite $\alpha \circ \alpha : C(V) \to C(V)$
is now an algebra homomorphism restricting to V as the identity and
hence coinciding on the whole of $C(V)$ with the identity. Thus α is in
fact an antiautomorphism of $C(V)$: indeed, it is the unique antiautomorphism of $C(V)$ that fixes V pointwise. We shall refer to α as the
main antiautomorphism of the complex Clifford algebra. Incidentally, α
arises also as follows: reversal of all tensor products defines an antiautomorphism of the tensor algebra $T(V)$ stabilizing the ideal $I(V)$ and α
is the antiautomorphism induced on the quotient $T(V)/I(V) = C(V)$.

Let us denote by $\overline{C(V)}$ the algebra conjugate to $C(V)$: thus, $\overline{C(V)}$
is $C(V)$ as a set, with precisely the same ring structure but with con-
jugated scalar multiplication, so that the identity map $C(V) \to \overline{C(V)}$ is
an antilinear ring isomorphism. The canonical inclusion $V \to \overline{C(V)}$ being a Clifford map, the universal mapping property for $C(V)$ provides a
unique algebra homomorphism κ from $C(V)$ to $\overline{C(V)}$ restricting to V as
the identity; of course, we may view κ as an antilinear ring homomorphism from $C(V)$ to itself. Being an algebra homomorphism restricting
to V as the identity, the composite $\kappa \circ \kappa : C(V) \to C(V)$ is the identity
on all of $C(V)$. Thus, κ is an antilinear ring automorphism of $C(V)$:
in fact, it is the unique such fixing V pointwise. We shall refer to κ as
the *main conjugation* of the complex Clifford algebra, often writing $\overline{a}$ in
place of $\kappa(a)$ when $a \in C(V)$. Incidentally, the conjugation of $V^{\mathbb{C}}$ pointwise fixing V extends functorially to a conjugation of $T(V)$ stabilizing
$I(V)$; the main conjugation of $C(V)$ is the induced map on the quotient
$T(V)/I(V)$.

Now the main antiautomorphism α and the main conjugation κ commute; their product is the unique antilinear antiautomorphism of $C(V)$
restricting to V as the identity. Thus, $\alpha \circ \kappa = \kappa \circ \alpha$ is an involution or
adjoint operation: we shall call it the *main involution* of the complex

Clifford algebra and shall denote it by a star, so that if $a \in C(V)$ then

$$a^* = \alpha(\bar{a}) = \overline{\alpha(a)}.$$

In this way, $C(V)$ naturally becomes an algebra with involution, or involutive algebra. As such, it satisfies a further universal mapping property, the statement of which requires a definition: if B is an involutive unital associative complex algebra, then the Clifford map $f : V \to B$ is *self-adjoint* in case $f(v)^* = f(v)$ whenever $v \in V$.

Theorem 1.1.3 *If $f : V \to B$ is a self-adjoint Clifford map then the unique algebra map $F : C(V) \to B$ such that $F \mid V = f$ is involution-preserving.*

Proof Simply note that the set $\{a \in C(V) : F(a)^* = F(a^*)\}$ is a subalgebra of $C(V)$ containing V and recall that V generates $C(V)$ as a complex algebra. $\qquad \square$

In this regard, it should be noted that if $g \in O(V)$ then the Bogoliubov automorphism θ_g is involution-preserving and hence an automorphism of $C(V)$ as an involutive algebra; moreover, θ_g commutes with the grading automorphism, the main antiautomorphism and the main conjugation.

After these remarks on complex Clifford algebras in general, we now pay more particular attention to the finite-dimensional situation. Thus, let the real inner product space V be finite-dimensional with $\{v_1, \ldots, v_m\}$ as a specific orthonormal basis. It is notationally convenient to write $\mathbf{m}$ in place of $\{1, \ldots, m\}$. If $S = \{s_1 < \ldots < s_p\}$ is a nonempty subset of $\mathbf{m}$ then we shall put

$$v_S = v_{s_1} \ldots v_{s_p}$$

with the product formed in $C(V)$. By convention, we shall associate the multiplicative identity of $C(V)$ to the empty index: $v_\emptyset = \mathbf{1}$. Notice that v_S is a unitary element of the involutive algebra $C(V)$ whenever $S \subset \mathbf{m}$: on the one hand, vectors in V are self-adjoint in being fixed by the main involution; on the other hand, unit vectors in V have square $\mathbf{1}$ on account of the Clifford property. It turns out that $\{v_S : S \subset \mathbf{m}\}$ is a basis for $C(V)$ as a complex vector space, whence $C(V)$ has complex dimension $2^{|\mathbf{m}|} = 2^m$. Our route towards establishing this fact lies by way of properties of the elements $\{v_S : S \subset \mathbf{m}\}$ that prove rather useful in probing further the structure of $C(V)$.

First, let S and T be subsets of $\mathbf{m}$ having cardinalities $|S|$ and $|T|$ respectively. Repeated application of the Clifford relations shows that

$$v_T v_S = (-1)^{|S||T|} v_S v_T$$

whenever S and T are disjoint. In general, we have the following result.

Theorem 1.1.4   *If $S, T \subset \mathbf{m}$ then $v_T v_S = (-1)^{|S|\,|T|+|S\cap T|} v_S v_T$.*

Proof Put $R = S \cap T$, $S' = S - R$ and $T' = T - R$; indicate cardinalities by the corresponding lower case letters. Note that

$$v_S = \sigma v_R v_{S'}, \qquad v_T = \tau v_R v_{T'}$$

where the signs $\sigma, \tau \in \{+1, -1\}$ arise from the Clifford relations as a result of reordering. The special case recorded before the theorem implies that

$$\tau \sigma v_T v_S = v_R v_{T'} v_R v_{S'}$$
$$= (-1)^{rt'} v_R v_R v_{T'} v_{S'}$$
$$= (-1)^{rt'+s'r+s't'} v_R v_{S'} v_R v_{T'}$$
$$= (-1)^{rt'+s'r+s't'} \sigma \tau v_S v_T.$$

Moreover,

$$rt' + s'r + s't' = (r + s')(r + t') - r^2$$

is congruent to

$$(r + s')(r + t') + r = st + r$$

modulo 2. Consequently,

$$v_T v_S = (-1)^{st+r} v_S v_T$$

and the proof is complete.   □

We pause to reformulate this result and to consider some special cases. Recall that if $T \subset \mathbf{m}$ then $v_T \in C(V)$ is unitary. As a consequence, the identity just established can be reformulated as saying that if S and T are subsets of $\mathbf{m}$ then

$$v_T v_S v_T^* = (-1)^{|S|\,|T|+|S\cap T|} v_S.$$

In particular, if $S \subset \mathbf{m}$ and if $j \in \mathbf{m}$ then

$$v_j v_S v_j = (-1)^{|S|+|S\cap j|} v_S.$$

More particularly still, if $|S|$ is even then

$$v_j v_S v_j = \begin{cases} +v_S & (j \notin S) \\ -v_S & (j \in S) \end{cases}$$

whilst if $|S|$ is odd then

$$v_j v_S v_j = \begin{cases} -v_S & (j \notin S) \\ +v_S & (j \in S) \end{cases}.$$

This reformulation and these special cases turn out to be particularly valuable in our analysis of the complex Clifford algebra.

The remaining property of the vectors $\{v_S : S \subset \mathbf{m}\}$ will not be needed in establishing that these vectors form a basis for $C(V)$ but it is important for other reasons and is conveniently disposed of at this point. In order to state the property, we require some notation. For subsets $S \subset \mathbf{m}$ and $T \subset \mathbf{m}$ we denote their *symmetric difference* by $S\Delta T$ as usual, so that

$$S\Delta T = (S - T) \cup (T - S);$$

in addition, we denote by $\varepsilon(S, T)$ the sign $(-1)^k$ where k is the cardinality of the set

$$\{(s, t) : s > t\} \subset S \times T.$$

Theorem 1.1.5 *If $S, T \subset \mathbf{m}$ then $v_S v_T = \varepsilon(S, T)v_{S\Delta T}$.*

Proof Let $S = \{s_1 < \ldots < s_p\}$ and $T = \{t_1 < \ldots < t_q\}$. For $j = 1, \ldots, p$ let k_j denote the cardinality of the set $\{t : s_j > t\} \subset T$ so that $\varepsilon(S, T) = (-1)^k$ where $k = k_1 + \ldots + k_p$. Making repeated use of the Clifford relations,

$$\begin{aligned}
v_S v_T &= v_{s_1} \ldots v_{s_p} v_{t_1} \ldots v_{t_q} \\
&= (-1)^{k_1} \ldots (-1)^{k_p} v_{S\Delta T} \\
&= (-1)^k v_{S\Delta T} \\
&= \varepsilon(S, T)v_{S\Delta T}
\end{aligned}$$

since $v_j^2 = \mathbf{1}$ for $j \in \mathbf{m}$ and since to arrange the elements of the set $S\Delta T$ in increasing order we must move s_r past each $t \in T$ with $s_r > t$ for $r = p, \ldots, 1$ (in that order). $\qquad\square$

Actually, the use to which we shall put this result only calls for the weaker result that if $S, T \subset \mathbf{m}$ then

$$v_S v_T = \pm v_{S\Delta T}$$

and does not require a determination of the sign.

We are now able to establish the advertised fact that $\{v_S : S \subset \mathbf{m}\}$ is a basis for $C(V)$. That $\{v_S : S \subset \mathbf{m}\}$ spans $C(V)$ is almost immediate from Theorem 1.1.1, according to which the algebra $C(V)$ is generated by its subspace V; all we need note in addition is that the Clifford relations permit the reduction of any finite product from $\{v_1, \ldots, v_m\}$ to one of the form v_S for some $S \subset \mathbf{m}$. Of course, this already implies that $C(V)$ is finite-dimensional. To see that $\{v_S : S \subset \mathbf{m}\}$ is linearly independent, suppose

$$\sum_{S \subset \mathbf{m}} \mu_S v_S = 0$$

to be a nontrivial relation involving as few nonzero coefficients as possible. This minimality and an application of the idempotent operators $\frac{1}{2}(I + \gamma)$ and $\frac{1}{2}(I - \gamma)$ together show at once that the indices $S \subset \mathbf{m}$ for which $\mu_S \neq 0$ all have the same parity: either $|S|$ is even whenever $\mu_S \neq 0$ or $|S|$ is odd whenever $\mu_S \neq 0$. Now hypothesize that the relation involves (at least) two nonzero coefficients; select j in the symmetric difference of the corresponding pair of indices in $\mathbf{m}$. From the first special case following Theorem 1.1.4 we deduce that

$$0 = v_j \Big(\sum_{S \subset \mathbf{m}} \mu_S v_S \Big) v_j$$

$$= \sum_{S \subset \mathbf{m}} (-1)^{|S| + |S \cap j|} \mu_S v_S$$

whence

$$0 = \sum_{S \subset \mathbf{m}} (-1)^{|S \cap j|} \mu_S v_S$$

since the sign $(-1)^{|S|}$ is constant over $\{S \subset \mathbf{m} : \mu_S \neq 0\}$. By hypothesis, addition of this relation to the original will result in a nontrivial relation having fewer nonzero coefficients, a patent absurdity. The supposed nontrivial relation cannot have just one nonzero coefficient since v_S is invertible whenever $S \subset \mathbf{m}$. Thus, the supposed nontrivial relation among the vectors $\{v_S : S \subset \mathbf{m}\}$ is nonexistent, so that $\{v_S : S \subset \mathbf{m}\}$ is indeed linearly independent. Of course, it now follows that $C(V)$ has complex dimension $2^{|\mathbf{m}|} = 2^m$.

Theorem 1.1.6 *If $\{v_1, \ldots, v_m\}$ is an orthonormal basis for V then $\{v_S : S \subset \mathbf{m}\}$ is a basis for $C(V)$ so that*

$$\dim_{\mathbb{C}} C(V) = 2^{\dim_{\mathbb{R}} V}.$$

$\square$

A little later, we shall offer an alternative proof that the vectors $\{v_S : S \subset \mathbf{m}\}$ are linearly independent: Theorem 1.1.9 states that $C(V)$ carries a natural positive-definite Hermitian inner product, relative to which $\{v_S : S \subset \mathbf{m}\}$ is an orthonormal basis. Our construction of this natural inner product will be performed with the aid of another natural structure carried by the complex Clifford algebra: namely, a normalized even central linear functional which we call its trace. Our handling of this trace is facilitated by having access to the left regular representation of the complex Clifford algebra. In order not to interrupt the development of the trace, it is convenient to present a brief introduction to the left regular representation at this juncture.

As with any algebra, the complex Clifford algebra $C(V)$ acts on itself by left multiplication. The *left regular representation*

$$\lambda : C(V) \to \operatorname{End} C(V)$$

is defined by the rule

$$\lambda(a)\zeta = a\zeta$$

for a and ζ in $C(V)$. Notice that if $a \in C^+(V)$ then $\lambda(a)$ leaves invariant each of $C^+(V)$ and $C^-(V)$. Notice also that if $a \in C^-(V)$ then $\lambda(a)$ maps $C^+(V)$ to $C^-(V)$ and maps $C^-(V)$ to $C^+(V)$. In particular, if $a \in C^-(V)$ then $\lambda(a)$ has block form

$$\lambda(a) = \begin{bmatrix} 0 & * \\ * & 0 \end{bmatrix}$$

relative to the direct sum decomposition

$$C(V) = C^+(V) \oplus C^-(V)$$

into eigenspaces of the grading automorphism. This is all that we require of the left regular representation for now; we shall return to examine it in greater detail later.

Now, let $\tau : C(V) \to \mathbb{C}$ be a linear functional on the complex Clifford algebra. We say that τ is *normalized* if and only if $\tau(\mathbf{1}) = 1$ and is *central* if and only if it satisfies

$$a, b \in C(V) \quad \Rightarrow \quad \tau(ba) = \tau(ab).$$

We say that τ is *even* if and only if it satisfies either of the following equivalent conditions: that $\tau = \tau \circ \gamma$ is invariant under the grading automorphism; that τ is identically zero on $C^-(V)$. In these terms, we claim that $C(V)$ carries a unique normalized even central linear functional.

Attending first to uniqueness, let $\tau : C(V) \to \mathbb{C}$ be a normalized even central linear functional. Normalization of τ means that $\tau(v_\emptyset) = \tau(\mathbf{1}) = 1$ whilst the even nature of τ ensures that if $S \subset \mathbf{m}$ and $|S|$ is odd then $\tau(v_S) = 0$. Now let $S \subset \mathbf{m}$ be such that $|S| > 0$ is even and let $j \in S$ be minimal, so that if $T = S - \{j\}$ then $v_S = v_j v_T$. Repeated use of the Clifford relations yields

$$v_j v_T = (-1)^{|T|} v_T v_j = -v_T v_j$$

since $|T|$ is odd; the central nature of τ consequently forces $\tau(v_S) = \tau(v_j v_T) = 0$. Thus τ is indeed unique, being given necessarily by the prescription

$$S \subset \mathbf{m} \quad \Rightarrow \quad \tau(v_S) = \begin{cases} 1 & (S = \emptyset) \\ 0 & (S \neq \emptyset). \end{cases}$$

Knowing from Theorem 1.1.6 that $\{v_S : S \subset \mathbf{m}\}$ is a basis for the complex Clifford algebra, it is not difficult to verify that the linear map

$\tau : C(V) \to \mathbb{C}$ uniquely determined by the prescription above is indeed a normalized even central linear functional: in fact, only centrality requires any proof and this follows from the observation that if $S, T \subset \mathbf{m}$ are unequal then $S \bigtriangleup T \neq \emptyset$ so that $\tau(v_S v_T) = \tau(\pm v_{S \bigtriangleup T}) = 0$ in view of Theorem 1.1.5.

Existence of τ can be established independently of these considerations, using instead the left regular representation λ. Recall the elementary consequence of Theorem 1.1.1 that $C(V)$ has $\{v_S : S \subset \mathbf{m}\}$ as a spanning set and is therefore finite-dimensional. This circumstance permits us to define a linear functional τ on $C(V)$ by

$$a \in C(V) \quad \Rightarrow \quad \tau(a) = \frac{\mathrm{Tr}\,(\lambda(a))}{\dim C(V)}$$

where Tr denotes trace as usual. The linear functional τ is normalized by construction, since $\lambda(\mathbf{1})$ is the identity operator on $C(V)$. The block formula

$$a \in C^-(V) \quad \Rightarrow \quad \lambda(a) = \begin{bmatrix} 0 & * \\ * & 0 \end{bmatrix}$$

makes it plain that τ vanishes on $C^-(V)$ and hence is even. Finally, τ is central since the same is true of the usual trace on linear endomorphisms.

In summary, we have justified our claim to the effect that $C(V)$ carries a unique normalized even central linear functional; we formulate our findings as follows.

Theorem 1.1.7 *$C(V)$ possesses a unique normalized even central linear functional, its trace τ defined by*

$$S \subset \mathbf{m} \quad \Rightarrow \quad \tau(v_S) = \begin{cases} 1 & (S = \emptyset) \\ 0 & (S \neq \emptyset). \end{cases}$$

$\square$

Thus the *trace* τ on $C(V)$ is given by

$$\tau\Big(\sum_{S \subset \mathbf{m}} \mu_S v_S\Big) = \mu_\emptyset$$

for any complex scalars $\{\mu_S : S \subset \mathbf{m}\}$. The following *Hermitian* property of τ can be verified from this formula; however, we present an alternative argument that is independent of basis.

Theorem 1.1.8 *If $a \in C(V)$ then $\tau(a^*) = \overline{\tau(a)}$.*

Proof We define a linear functional σ on $C(V)$ by stipulating that

$\sigma(a) = \overline{\tau(a^*)}$ whenever $a \in C(V)$. It is a matter of routine verification that σ is normalized, even and central; hence $\sigma = \tau$ by uniqueness of the trace. $\qquad\square$

The complex Clifford algebra will now be seen to carry a canonical positive-definite Hermitian inner product. Explicitly, for $\xi, \eta \in C(V)$ we put

$$\langle \xi \mid \eta \rangle = \tau(\eta^*\xi).$$

The form $\langle \cdot \mid \cdot \rangle$ so defined is quite plainly sesquilinear; as a result of Theorem 1.1.8 it is actually Hermitian, since if $\xi, \eta \in C(V)$ then

$$\langle \eta \mid \xi \rangle = \tau(\xi^*\eta) = \overline{\tau(\eta^*\xi)} = \overline{\langle \xi \mid \eta \rangle}.$$

Finally, we contend that if $S, T \subset \mathbf{m}$ then

$$\langle v_S \mid v_T \rangle = \begin{cases} 1 & (S = T) \\ 0 & (S \neq T) \end{cases}$$

as a result of which $\langle \cdot \mid \cdot \rangle$ is positive-definite; indeed it follows that $\{v_S : S \subset \mathbf{m}\}$ is an orthonormal basis for $C(V)$. In the first place, if $S \subset \mathbf{m}$ then v_S is unitary, so

$$\langle v_S \mid v_S \rangle = \tau(v_S^* v_S) = \tau(\mathbf{1}) = 1.$$

In case the indices $S, T \subset \mathbf{m}$ are unequal, their symmetric difference $S \,\Delta\, T = T \,\Delta\, S$ is nonempty, whence from Theorem 1.1.5 it follows that

$$\langle v_S \mid v_T \rangle = \tau(v_T^* v_S) = \pm\tau(v_T v_S)$$
$$= \pm\tau(v_{T \,\Delta\, S}) = 0$$

since

$$v_T^* = \alpha(v_T) = (-1)^{\frac{1}{2}|T|\,(|T|-1)}\, v_T$$

on account of the Clifford relations.

Theorem 1.1.9 $\{v_S : S \subset \mathbf{m}\}$ *is an orthonormal basis for* $C(V)$ *relative to the canonical positive-definite Hermitian inner product defined by*

$$\xi, \eta \in C(V) \quad \Rightarrow \quad \langle \xi \mid \eta \rangle = \tau(\eta^*\xi).$$

$\qquad\square$

In particular, note that the norm $\| \cdot \|$ engendered by $\langle \cdot \mid \cdot \rangle$ on $C(V)$ is given by

$$\left\| \sum_{S \subset \mathbf{m}} \mu_S v_S \right\|^2 = \sum_{S \subset \mathbf{m}} |\mu_S|^2$$

for any collection $\{\mu_S : S \subset \mathbf{m}\}$ of complex coefficients.

Much of the further structure of complex Clifford algebras in finite dimensions depends largely on whether the dimension of the underlying real inner product space is even or odd. The case of odd dimensions holds the least interest for us; we shall favour it with a few remarks at the close of this section. For now, we spend some time on the special properties of complex Clifford algebras over *even-dimensional* real inner product spaces.

Thus, let the dimension m of V be even and recall the fixed orthonormal basis $\{v_1, \ldots, v_m\}$ of V. For convenience, let us write

$$\omega = v_{\mathbf{m}} = v_1 \ldots v_m.$$

This element of $C(V)$ holds the key to a number of structural properties of the complex Clifford algebra. From the reformulation of Theorem 1.1.4 it follows that if S is any subset of $\mathbf{m}$ then

$$\omega v_S \omega^* = v_{\mathbf{m}} v_S v_{\mathbf{m}}^*$$
$$= (-1)^{|S| \, |\mathbf{m}| + |S \cap \mathbf{m}|} v_S$$
$$= (-1)^{|S|} v_S$$

since $S \subset \mathbf{m}$ and $|\mathbf{m}| = m$ is even. Thus if $S \subset \mathbf{m}$ then

$$\omega v_S \omega^* = \begin{cases} +v_S & (\ |S| \text{ even}) \\ -v_S & (\ |S| \text{ odd}). \end{cases}$$

Since $C(V)$ has $\{v_S : S \subset \mathbf{m}\}$ as a basis, it now follows by linearity that if $a \in C(V)$ then $\omega a \omega^* = \gamma(a)$. We have established the following result.

Theorem 1.1.10 *If $m = \dim V$ is even and if $\{v_1, \ldots, v_m\}$ is an orthonormal basis for V then $\omega = v_1 \ldots v_m$ satisfies*

$$a \in C(V) \quad \Rightarrow \quad \gamma(a) = \omega a \omega^*.$$

$\square$

Otherwise said, the grading automorphism γ of $C(V)$ is *inner*, being implemented by the unitary ω. This fact enables us to demonstrate that the algebra $C(V)$ is both *central* (in the sense that its centre comprises precisely all scalar multiples of the multiplicative identity) and *simple* (in that it lacks nontrivial bilateral ideals).

Taking up first the matter of $C(V)$ being central, let

$$a = \sum_{S \subset \mathbf{m}} \mu_S v_S$$

lie in the centre of $C(V)$. In the light of the fact that

$$\gamma(a) = \omega a \omega^* = a$$

we see clearly that $a \in C^+(V)$ is even. Now, if $j \in \mathbf{m}$ then it follows from Theorem 1.1.4 that

$$\sum_{S \subset \mathbf{m}} \mu_S v_S = a = v_j a v_j$$

$$= \sum_{S \subset \mathbf{m}} \mu_S v_j v_S v_j$$

$$= \sum_{S \subset \mathbf{m}} (-1)^{|S \cap j|} \mu_S v_S$$

since μ_S is zero when $|S|$ is odd. As a result, if $\mu_S \neq 0$ then $|S \cap j|$ is even and so of course $j \notin S$. In short, if μ_S is nonzero then $S = \emptyset$; this means that $a = \mu_\emptyset v_\emptyset$ is indeed a scalar multiple of $\mathbf{1} = v_\emptyset$.

Passing on to simplicity, let $D \subset C(V)$ be a nonzero bilateral ideal; we must show that $D = C(V)$. First of all, from

$$\gamma(D) = \omega D \omega^* = D$$

it follows that D is graded in the sense that $D = D^+ \oplus D^-$ where $D^\pm = D \cap C^\pm(V)$. Notice that D^+ must be nonzero, since D^+ and D^- are interchanged isomorphically under left multiplication by any unit vector from V. Let

$$d = \sum_{S \subset \mathbf{m}} \mu_S v_S \in D^+$$

be a nonzero element in D^+ having as few nonzero coefficients as possible. If $j \in \mathbf{m}$ then from Theorem 1.1.4 it follows that

$$D^+ \ni v_j d v_j = \sum_{S \subset \mathbf{m}} (-1)^{|S \cap j|} \mu_S v_S$$

whence by addition and subtraction of d it follows that D^+ contains each of $\Sigma\{\mu_S v_S : j \in S \subset \mathbf{m}\}$ and $\Sigma\{\mu_S v_S : j \notin S \subset \mathbf{m}\}$. One of these sums is d by minimality, whence we deduce that either $j \in \bigcap\{S \subset \mathbf{m} : \mu_S \neq 0\}$ or $j \notin \bigcup\{S \subset \mathbf{m} : \mu_S \neq 0\}$. In consequence, precisely one coefficient μ_S in d is nonzero, thus $D \supset D^+$ contains the unit $d = \mu_S v_S$ and so $D = C(V)$.

Theorem 1.1.11 *If V is even-dimensional then the complex Clifford algebra $C(V)$ is both central and simple.* $\qquad\square$

Incidentally, the fact that $C(V)$ is central implies that unitary elements implementing the grading automorphism are proportional to one another; in particular, unitary implementers that are real (in being fixed by the main conjugation) are unique up to sign. Thus, the real unitary element $\omega = v_1 \ldots v_m$ implementing γ is independent of the orthonormal basis $\{v_1, \ldots, v_m\}$ except for a sign. Of course, this can be seen otherwise: for example, the Clifford relations imply that the product

$v_1 \ldots v_m$ is multiplied by $\det g = \pm 1$ when $\{v_1, \ldots, v_m\}$ is replaced by its transform $\{gv_1, \ldots, gv_m\}$ under any $g \in O(V)$.

We record here just one more consequence of the fact that γ is an inner automorphism when V is even-dimensional.

Theorem 1.1.12 *If V is even-dimensional then τ is the unique normalized central linear functional on $C(V)$.*

Proof It is enough to demonstrate that a central linear functional $\sigma :$ $C(V) \to \mathbb{C}$ is automatically even. For this, note that if $a \in C(V)$ then

$$\sigma\big(\gamma(a)\big) = \sigma(\omega a \omega^*) = \sigma(a)$$

since σ is central and ω is unitary. Thus, σ is invariant under γ and so even, as required. $\qquad\qquad\square$

We now turn our attention towards infinite dimensions: thus, let V be an *infinite-dimensional* real inner product space. Write $\mathcal{F}(V)$ for the set of all finite-dimensional subspaces of V, write $\mathcal{F}^+(V)$ for the even-dimensional subspaces and write $\mathcal{F}^-(V)$ for the odd-dimensional subspaces. Note that each of the sets $\mathcal{F}(V)$, $\mathcal{F}^+(V)$ and $\mathcal{F}^-(V)$ is directed under inclusion: for example, if $M, N \in \mathcal{F}(V)$ then $M + N$ lies in $\mathcal{F}(V)$ and contains both M and N. Note also that if $M \in \mathcal{F}(V)$ then $V = M \oplus M^\perp$ where the *orthocomplement* $M^\perp$ of M is the subspace of V defined by

$$M^\perp = \{v \in V : (M \mid v) = 0\};$$

indeed, if $\{v_1, \ldots, v_m\}$ is an orthonormal basis for M and if $v \in V$ then

$$\sum_{j \in \mathbf{m}} (v \mid v_j) v_j \in M$$

and

$$v - \sum_{j \in \mathbf{m}} (v \mid v_j) v_j \in M^\perp.$$

Our approach to an understanding of $C(V)$ is via finite-dimensional approximation: we probe the structure of $C(V)$ using the complex Clifford algebras over finite-dimensional subspaces of V as tools. With this in mind, let $M \in \mathcal{F}(V)$ be a finite-dimensional subspace of V and note that as an isometry, the inclusion $M \to V$ induces an algebra map $C(M) \to C(V)$ by universality. Upon reconsideration, the argument leading up to Theorem 1.1.6 actually establishes that if $\{v_1, \ldots, v_m\}$ is an orthonormal basis for M then the vectors $\{v_S : S \subset \mathbf{m}\}$ remain linearly independent when placed in $C(V)$. Thus, the canonical algebra map

$C(M) \to C(V)$ is injective and so allows us to identify $C(M)$ with its image, the subalgebra of $C(V)$ generated by M. Henceforth, we shall make this natural and convenient identification without comment. It is harmless, of course: after all, complex Clifford algebras are unique up to canonical isomorphisms.

Theorem 1.1.13 $C(V) = \bigcup\{C(M) : M \in \mathcal{F}(V)\}$.

Proof Recall from Theorem 1.1.1 that $C(V)$ is generated by V. This means that each $a \in C(V)$ may be expressed as a finite sum of finite products of vectors taken from V. If we take such an expression and let M be the linear span of the finite set of vectors involved, then clearly $a \in C(M)$. The theorem follows. $\square$

Note that if $a \in C(V)$ and if $M \in \mathcal{F}(V)$ is such that $a \in C(M)$ then we may suppose that the dimension of M is even or odd as we please, by the simple expedient of enlargement by one dimension if necessary. It follows that

$$C(V) = \bigcup\{C(M) : M \in \mathcal{F}^+(V)\}$$

and

$$C(V) = \bigcup\{C(M) : M \in \mathcal{F}^-(V)\}.$$

Similar remarks apply to the even complex Clifford algebra $C^+(V)$: this is the union of its subalgebras $C^+(M)$ as M runs over $\mathcal{F}(V)$, $\mathcal{F}^+(V)$ or $\mathcal{F}^-(V)$.

We make immediate use of these observations by showing that $C(V)$ inherits the properties of being central and simple.

Theorem 1.1.14 *If V is infinite-dimensional then the complex Clifford algebra $C(V)$ is both central and simple.*

Proof Let a lie in the centre of $C(V)$ and choose $M \in \mathcal{F}^+(V)$ such that $a \in C(M)$. Theorem 1.1.11 tells us that $C(M)$ has scalar centre, whence a is scalar. This shows that the algebra $C(V)$ is central. Now let $D \subset C(V)$ be a nonzero bilateral ideal of which d is a nonzero element and pick $N \in \mathcal{F}^+(V)$ so that $d \in C(N)$. Theorem 1.1.11 tells us that $C(N)$ is simple and hence equals its nonzero ideal $D \cap C(N)$, whence $D \supset C(N) \ni 1$ and therefore $D = C(V)$. This shows that the algebra $C(V)$ is simple. $\square$

The infinite-dimensional complex Clifford algebra $C(V)$ also carries a canonical trace. Of course, this trace cannot be constructed by normalizing the operator trace on the left regular representation; instead, we fashion it by matching the traces on complex Clifford algebras over finite-dimensional subspaces of V. The interests of clarity will be promoted by our agreeing here that if $M \in \mathcal{F}(V)$ then the canonical trace on $C(M)$ will be written as τ_M rather than as τ alone. Now, given $a \in C(V)$ we define $\tau(a)$ as follows. First of all, there exists a finite-dimensional subspace $M \in \mathcal{F}(V)$ of V such that $a \in C(M)$; in this case, we put $\tau(a) = \tau_M(a)$. Next, if also $N \in \mathcal{F}(V)$ with $a \in C(N)$ then the canonical trace on $C(M + N)$ restricts to τ_M on $C(M)$ and τ_N on $C(N)$ by uniqueness, whence $\tau_M(a) = \tau_N(a)$ and $\tau(a)$ is unambiguously defined. That $\tau : C(V) \to \mathbb{C}$ so defined is a normalized even central linear functional is clear from the fact that $\tau \mid C(M) = \tau_M$ whenever M lies in the directed set $\mathcal{F}(V)$. As before, we call τ the *trace* of the infinite-dimensional complex Clifford algebra $C(V)$.

As was true of the trace on complex Clifford algebras over even-dimensional real inner product spaces, τ is actually unique as a normalized central linear functional on $C(V)$; again, we establish this by working up through even-dimensional subspaces of V.

Theorem 1.1.15 *If V is infinite-dimensional then τ is the unique normalized central linear functional on $C(V)$.*

Proof Let $\sigma : C(V) \to \mathbb{C}$ be a normalized central linear functional. If $M \in \mathcal{F}^+(V)$ then $\sigma \mid C(M)$ is a normalized central linear functional on $C(M)$ and therefore coincides with τ_M according to Theorem 1.1.12. The equality $\sigma = \tau$ now follows since $C(V)$ is the union of its subalgebras $C(M)$ as M runs over $\mathcal{F}^+(V)$. $\qquad\qquad\square$

Its uniqueness as a normalized (even) central linear functional on $C(V)$ ensures that τ is invariant under all automorphisms and antiautomorphisms of the complex Clifford algebra. As a matter of fact, rather weaker properties serve to distinguish the trace: for example, we claim that it is the unique normalized linear functional on $C(V)$ invariant under all Bogoliubov automorphisms in the sense that $\tau \circ \theta_g = \tau$ whenever $g \in O(V)$.

In preparation for a justification of this claim, let $v \in V$ be a unit vector and consider the inner automorphism of $C(V)$ given by

$$C(V) \to C(V) : a \mapsto vav.$$

The Clifford relations imply that this automorphism fixes v and sends each vector in the orthocomplement $v^\perp \subset V$ to its negative; it is therefore the Bogoliubov automorphism induced by minus orthogonal reflection in the hyperplane $v^\perp$ perpendicular to v.

For the claim itself, let $\sigma : C(V) \to \mathbb{C}$ be a normalized linear functional that is invariant under all Bogoliubov automorphisms. Our preparatory considerations above imply that if $v \in V$ is a unit vector and if $a \in C(V)$ then

$$\sigma(va) = \sigma(vvav) = \sigma(av);$$

of course, $\sigma(va) = \sigma(av)$ then holds for all $v \in V$ and all $a \in C(V)$ by linearity. Now the *centralizer*

$$\{b \in C(V) : a \in C(V) \quad \Rightarrow \quad \sigma(ba) = \sigma(ab)\}$$

of σ is a subalgebra of $C(V)$ containing V and is consequently all of $C(V)$ by virtue of Theorem 1.1.1. Having thus established that the normalized linear functional $\sigma : C(V) \to \mathbb{C}$ is central, Theorem 1.1.15 justifies our claim to the effect that $\sigma = \tau$.

We record our justified claim as follows.

Theorem 1.1.16 *If V is infinite-dimensional then τ is the unique normalized linear functional on $C(V)$ that is invariant under all Bogoliubov automorphisms.* $\qquad\square$

As was the case in finite dimensions, the trace τ on the infinite-dimensional complex Clifford algebra $C(V)$ is Hermitian in the sense that

$$a \in C(V) \quad \Rightarrow \quad \tau(a^*) = \overline{\tau(a)}.$$

This readily follows from Theorem 1.1.8 upon invoking the fact that $\tau \mid C(M) = \tau_M$ whenever $M \in \mathcal{F}(V)$. In consequence, the sesquilinear form $\langle \cdot \mid \cdot \rangle$ defined on $C(V)$ by the prescription

$$\xi, \eta \in C(V) \quad \Rightarrow \quad \langle \xi \mid \eta \rangle = \tau(\eta^*\xi)$$

is actually Hermitian; it is moreover positive-definite, since it restricts to the form of Theorem 1.1.9 on $C(M)$ whenever $M \in \mathcal{F}(V)$. We remark that the inner product $\langle \cdot \mid \cdot \rangle$ is invariant under all automorphisms of $C(V)$ as an involutive algebra, since τ is invariant under all algebra automorphisms of $C(V)$. In particular, if $g \in O(V)$ and if $\xi, \eta \in C(V)$ then $\langle \theta_g\xi \mid \theta_g\eta \rangle = \langle \xi \mid \eta \rangle$.

Theorem 1.1.17 *The positive-definite Hermitian inner product defined canonically on $C(V)$ by the rule*

$$\xi, \eta \in C(V) \quad \Rightarrow \quad \langle \xi \mid \eta \rangle = \tau(\eta^*\xi)$$

is invariant under all automorphisms of $C(V)$ as an involutive algebra.

□

Thus far, the even complex Clifford algebra $C^+(V)$ has largely been neglected; it is now appropriate to remedy this state of affairs, for V of any dimension. To do so, let $l \in V$ be a unit vector and let L be its one-dimensional linear span with $L^\perp$ the orthocomplementary hyperplane. We claim that the even complex Clifford algebra $C^+(V)$ and the complex Clifford algebra $C(L^\perp)$ are isomorphic as involutive algebras.

Indeed, let us define a plainly linear map $\Phi : C(L^\perp) \to C^+(V)$ by requiring that Φ be the identity on $C^+(L^\perp) \subset C^+(V)$ and that Φ send $a \in C^-(L^\perp)$ to $ila \in C^+(V)$. If each of a and b lies in $C^-(L^\perp)$ then

$$\Phi(a)\Phi(b) = ilailb = -lalb$$

$$= llab = ab$$

$$= \Phi(ab)$$

since l anticommutes with elements of $C^-(L^\perp)$ on account of the Clifford relations as in Theorem 1.1.4. Since l commutes with elements of $C^+(L^\perp)$ for similar reasons, it is also true that $\Phi(a)\Phi(b) = \Phi(ab)$ in case the elements a and b of $C(L^\perp)$ have opposite parity in the sense that one is even and the other odd. It should now be clear that Φ is an algebra homomorphism. To see that Φ is surjective, let $a \in C^+(V)$ and suppose that $a \in C^+(M)$ for some $M \in \mathcal{F}(V)$ containing l without loss. Extend $l = v_1$ to an orthonormal basis $\{v_1, \ldots, v_m\}$ for M and let

$$a = \sum_{S \subset \mathbf{m}} \mu_S v_S = a' + a''$$

where

$$a' = \sum \{\mu_S v_S : 1 \notin S\} \in C^+(L^\perp)$$

and

$$a'' = \sum \{\mu_S v_S : 1 \in S\} \in l \cdot C^-(L^\perp).$$

It follows that

$$a = \Phi(a' - ila'')$$

whence Φ is surjective. That Φ is also injective follows from a dimension count in finite dimensions and from the simplicity of $C(L^\perp)$ in infinite dimensions. Finally, the algebra isomorphism Φ is involution-preserving since elements of $C^-(L^\perp)$ anticommute with l.

Theorem 1.1.18 *Let V have any dimension and let $l \in V$ be a unit vector with linear span L. The map*

$$\Phi : C(L^\perp) \to C^+(V)$$

given by

$$a \in C^+(L^\perp) \quad \Rightarrow \quad \Phi(a) = a$$
$$a \in C^-(L^\perp) \quad \Rightarrow \quad \Phi(a) = ila$$

is an isomorphism of involutive algebras. □

We remark incidentally that the isomorphism Φ is in fact induced from the self-adjoint Clifford map

$$\phi : L^\perp \to C^+(V) : v \mapsto ilv$$

by the universal mapping property expressed in Theorem 1.1.3.

The theorem just established has important consequences, in particular when V is infinite-dimensional. In this case, if $l \in V$ is a unit vector having L as its linear span then of course $L^\perp$ is also infinite-dimensional. It follows from Theorem 1.1.14 that the complex Clifford algebra $C(L^\perp)$ is central and simple; the even complex Clifford algebra $C^+(V)$ inherits both of these properties by reason of Theorem 1.1.18.

Theorem 1.1.19 *If V is infinite-dimensional then the even complex Clifford algebra $C^+(V)$ is both central and simple.* □

In turn, this circumstance allows us to answer rather easily the question of whether or not the grading automorphism of $C(V)$ is inner when V is infinite-dimensional.

Theorem 1.1.20 *If V is infinite-dimensional then the grading automorphism γ of $C(V)$ is not inner.*

Proof Suppose γ to be inner and let it be implemented by the unit $u \in C(V)$ in the sense that $\gamma(a) = uau^{-1}$ whenever $a \in C(V)$. In particular, $\gamma(u) = uuu^{-1} = u$ so that u lies in $C^+(V)$; moreover, if $a \in C^+(V)$ then $ua = uau^{-1}u = \gamma(a)u = au$ so that u actually lies in the centre of $C^+(V)$. Since $C^+(V)$ has scalar centre it follows that u is a scalar and so cannot possibly implement γ. □

Recall that γ is the Bogoliubov automorphism of $C(V)$ restricting to V as minus the identity. We shall consider Bogoliubov automorphisms of $C(V)$ generally in Section 4.1, presenting a complete characterization of those that are inner.

Finally, we close this opening section by remarking upon some of the special properties of the complex Clifford algebra $C(V)$ when the real inner product space V is *odd-dimensional*. As the odd-dimensional case

is not of the greatest interest to us, we shall be brief. First of all, the
product $\omega = v_1 \ldots v_m$ of vectors in an orthonormal basis $\{v_1, \ldots, v_m\}$
for V actually lies in the centre of $C(V)$ by Theorem 1.1.4. This means
that the algebra $C(V)$ is not central: in fact, an argument only slightly
more elaborate than that for the first part of Theorem 1.1.11 shows that
the centre of $C(V)$ is the complex plane spanned by $\mathbf{1}$ and ω. We remark
that this centre contains a pair of odd elements with square $\mathbf{1}$: indeed,
$(\pm\omega)^2 = \mathbf{1}$ if $m \equiv 1 \pmod 4$ and $(\pm\mathrm{i}\omega)^2 = \mathbf{1}$ if $m \equiv 3 \pmod 4$. As a
consequence, $C(V)$ contains a pair of nontrivial central idempotents with
sum $\mathbf{1}$ and vanishing product: specifically, $\frac{1}{2}(\mathbf{1} \pm \omega)$ if $m \equiv 1 \pmod 4$
and $\frac{1}{2}(\mathbf{1} \pm \mathrm{i}\omega)$ if $m \equiv 3 \pmod 4$. Of course, $C(V)$ is now the direct sum of
the ideals generated by these idempotents, whence $C(V)$ is not simple
and Theorem 1.1.11 fails in its second part also. However, the even
complex Clifford algebra $C^+(V)$ is both central and simple: this follows
from Theorem 1.1.18 as in Theorem 1.1.19, since hyperplanes in V are of
even dimension and hence have central simple complex Clifford algebras
according to Theorem 1.1.11. As in Theorem 1.1.20, this circumstance
implies that the grading automorphism γ of $C(V)$ is not inner. Lastly,
$C(V)$ has an affine complex line of normalized central linear functionals:
if $\mu \in \mathbb{C}$ then

$$\tau_\mu : C(V) \to \mathbb{C} : a \mapsto \tau(a + \mu\omega a)$$

is such a functional, which reduces to the (even) trace $\tau = \tau_0$ when it
takes equal values on the two nontrivial central idempotents.

1.2 C^* Clifford algebras

To a given real inner product space V of arbitrary dimension, we have
associated its complex Clifford algebra $C(V)$: among other things, this
is a unital associative complex algebra with a canonical involution. Our
purpose in the present section is to introduce and study what is essentially the enveloping C^* algebra of $C(V)$: we shall call it the C^* Clifford
algebra of V and suggestively denote it by $C[V]$.

Thus, let V be a real inner product space of any dimension. Let $\mathrm{Rep}\,V$
signify the collection of all *star-homomorphisms* (that is, involution-
preserving homomorphisms) from $C(V)$ to arbitrary C^* algebras. For
our purposes, little harm would result were we to restrict attention to
star-representations of $C(V)$ on complex Hilbert spaces. We would like
to define a norm $\| \cdot \|_\infty$ on $C(V)$ by prescribing that if $a \in C(V)$ then

$$\|a\|_\infty = \sup\{ \|\pi(a)\| : \pi \in \mathrm{Rep}\,V \}$$

taking the supremum of the (operator) norms $\|\pi(a)\|$ for $\pi \in \mathrm{Rep}\,V$.

Were this legitimate, the completion of $C(V)$ in the norm $\| \cdot \|_\infty$ would then be a C^* algebra, which we could take for $C[V]$. There are two potential obstacles to legitimacy: on the one hand, it is not yet clear that the supremum defining $\| \cdot \|_\infty$ is always finite; on the other, it is not yet clear that the supremum $\|a\|_\infty$ is nonzero when $a \in C(V)$ is nonzero. We face each of these potential obstacles in turn.

Taking first the finiteness of the supremum, let us suppose that $a \in C(V)$. If V is infinite-dimensional, then Theorem 1.1.13 presents us with a finite-dimensional subspace $M \in \mathcal{F}(V)$ of V such that $a \in C(M)$; if V is finite-dimensional, then put $M = V$. In either case, let $\{v_1, \ldots, v_m\}$ be an orthonormal basis for M and let

$$a = \sum_{S \subset \mathbf{m}} \mu_S v_S.$$

Now, let $\pi \in \mathrm{Rep}\, V$ be any nonzero star-homomorphism from $C(V)$ to a C^* algebra B. If $S \subset \mathbf{m}$ then v_S is unitary in $C(V)$ so that $\pi(v_S)$ is unitary in B: in particular,

$$\|\pi(v_S)\|^2 = \|\pi(v_S)^*\pi(v_S)\|$$
$$= \|\pi(v_S^* v_S)\|$$
$$= \|\pi(\mathbf{1})\| = 1.$$

It follows that

$$\|\pi(a)\| = \|\sum_{S \subset \mathbf{m}} \mu_S \pi(v_S)\|$$
$$\leq \sum_{S \subset \mathbf{m}} |\mu_S| \, \|\pi(v_S)\|$$
$$= \sum_{S \subset \mathbf{m}} |\mu_S|$$

independently of π. This justifies our defining $\|a\|_\infty$ as the supremum of the norms $\|\pi(a)\|$ for $\pi \in \mathrm{Rep}\, V$.

The question of whether $\|\pi(a)\|_\infty$ is nonzero when $a \in C(V)$ is nonzero is settled in the affirmative by the left regular representation, as we proceed to demonstrate. Recall that $C(V)$ is equipped with a canonical positive-definite Hermitian inner product $\langle \cdot \mid \cdot \rangle$ given by the prescription

$$\xi, \eta \in C(V) \quad \Rightarrow \quad \langle \xi \mid \eta \rangle = \tau(\eta^* \xi)$$

where τ denotes the canonical trace on the complex Clifford algebra. In the interests of both clarity and convenience we write H_τ for the inner product space that results when $C(V)$ is equipped with $\langle \cdot \mid \cdot \rangle$ and write $\mathbb{H}_\tau$ for its Hilbert space completion. We likewise let Ω stand for the multiplicative identity $\mathbf{1} \in C(V)$ when we wish to consider it as a distinguished vector in either H_τ or $\mathbb{H}_\tau$. The left regular representation is now

$$\lambda : C(V) \to \text{End } H_\tau$$

given by

$$a \in C(V),\ \zeta \in H_\tau \quad \Rightarrow \quad \lambda(a)\zeta = a\zeta$$

where multiplication takes place within the complex Clifford algebra.

Notice that λ is involution-preserving when $C(V)$ has its main involution and when adjoint operators on H_τ are defined as usual. Explicitly, if $a \in C(V)$ and if $\xi, \eta \in H_\tau$ then

$$\begin{aligned}
\langle \lambda(a)\xi \mid \eta \rangle &= \tau\big(\eta^*(a\xi)\big) \\
&= \tau\big((a^*\eta)^*\xi\big) \\
&= \langle \xi \mid \lambda(a^*)\eta \rangle.
\end{aligned}$$

When applied to λ, a repetition of our argument above that if $a \in C(V)$ then $\|a\|_\infty$ is finite now shows that $\lambda(a)$ is a bounded linear operator on H_τ. By continuous extension, λ now yields a star-representation

$$\lambda : C(V) \to B(\mathbb{H}_\tau)$$

of the complex Clifford algebra $C(V)$ on the canonical complex Hilbert space $\mathbb{H}_\tau$. We shall refer to this as either the *left regular representation* or the *trace representation* interchangeably. Note that the representation λ is faithful: indeed, if $a \in C(V)$ then $\lambda(a)\Omega = a$ where $\Omega \in H_\tau$ stands for $1 \in C(V)$ as agreed. Since $\lambda \in \text{Rep } V$ is faithful, it follows that $\|a\|_\infty \geq \|\lambda(a)\|$ is nonzero when $a \in C(V)$ is nonzero.

Having thus surmounted the potential obstacles to its legitimacy, the prescription

$$a \in C(V) \quad \Rightarrow \quad \|a\|_\infty = \sup\{\ \|\pi(a)\| : \pi \in \text{Rep } V\ \}$$

does indeed define a norm $\| \cdot \|_\infty$ on the complex Clifford algebra. This norm $\| \cdot \|_\infty$ satisfies the C^* property

$$a \in C(V) \quad \Rightarrow \quad \|a^*a\|_\infty = \|a\|_\infty^2$$

as follows from the fact that if $\pi \in \text{Rep } V$ then the seminorm $\| \cdot \|_\pi$ defined on $C(V)$ by

$$a \in C(V) \quad \Rightarrow \quad \|a\|_\pi = \|\pi(a)\|$$

also satisfies the C^* property. It follows that the completion of $C(V)$ relative to $\| \cdot \|_\infty$ is a C^* algebra. We call this completion the C^* *Clifford algebra* of V and denote it by $C[V]$.

This is as convenient a place as any to note that the oft suppressed canonical embedding $\phi : V \to C(V) \subset C[V]$ is isometric, sending the original norm $\| \cdot \|$ on V to the norm $\| \cdot \|_\infty$ on $C(V)$ defined above.

Theorem 1.2.1   *The canonical embedding $\phi : V \to C[V]$ is isometric.*

Proof Since ϕ is a self-adjoint Clifford map, it follows that if $v \in V$ then

$$\|\phi(v)\|_\infty^2 = \|\phi(v^*)\phi(v)\|_\infty = \|\phi(v)^2\|_\infty$$
$$= \left\| \|v\|^2 \mathbf{1} \right\|_\infty = \|v\|^2.$$

□

Suppressing ϕ once again, we may say that $\| \cdot \|_\infty$ is an extension to $C(V) \subset C[V]$ of the original norm $\| \cdot \|$ on V.

Of course, if V and hence $C(V)$ is finite-dimensional, then $C(V)$ is already complete relative to the norm $\| \cdot \|_\infty$. In this case, $C[V] = C(V)$ and we have little else to say concerning the structure of the C^* Clifford algebra beyond what was said in the preceding section. For this reason, we shall suppose throughout the remainder of the present section that the real inner product space V is *infinite-dimensional*. This supposition introduces technical simplifications: for example, since the algebra $C(V)$ is simple, if $\pi \in \operatorname{Rep} V$ is nonzero then π is injective and so $\| \cdot \|_\pi$ is actually a norm.

It transpires that our construction of the C^* Clifford algebra $C[V]$ is seemingly more complicated than necessary. In fact, it turns out that the norm $\| \cdot \|_\pi$ is independent of the nonzero star-homomorphism $\pi \in \operatorname{Rep} V$ and so coincides with $\| \cdot \|_\infty$ on $C(V)$. Thus, we could have constructed the C^* Clifford algebra of V by completing the image of $C(V)$ in the codomain of any nonzero star-homomorphism, such as the trace representation λ. This fact and many others follow from the fundamental circumstance that the C^* Clifford algebra $C[V]$ is *simple*.

Theorem 1.2.2 *The C^* Clifford algebra $C[V]$ is simple.*

Proof We shall be done once we show that any nonzero C^* algebra map π from $C[V]$ to a C^* algebra is isometric in the sense that it preserves norms. Let $M \in \mathcal{F}^+(V)$ be an even-dimensional subspace of V. The restriction of π to $C(M)$ is nonzero: indeed, its vanishing would imply that of $\pi(\mathbf{1})$ and therefore that of π itself. Recalling from Theorem 1.1.11 that $C(M)$ is simple, it follows that the restriction of π to $C(M)$ is actually isometric. The remarks following Theorem 1.1.13 now show π to be isometric on $C(V)$. Finally, π is isometric on $C[V]$ by continuity. □

Now, let $\pi \in \operatorname{Rep} V$ be a nonzero star-homomorphism from $C(V)$ to a C^* algebra B and denote by $C_\pi[V]$ the completion of $C(V)$ in the norm $\| \cdot \|_\pi$ defined by $\|a\|_\pi = \|\pi(a)\|$ for $a \in C(V)$ as before; of course,

$C_\pi[V]$ may be taken to be the uniform closure of the range of π in B. By its very construction, $\| \cdot \|_\infty$ dominates $\| \cdot \|_\pi$ on $C(V)$; as a direct consequence, the identity map on $C(V)$ extends by continuity to yield a C^* algebra map $T_\pi : C[V] \to C_\pi[V]$. The simplicity of $C[V]$ forces T_π to be injective and indeed isometric: in particular, if $a \in C(V)$ then $\|a\|_\infty = \|a\|_\pi$ as was announced earlier. Moreover, T_π is also surjective: on the one hand, the range of T_π is closed since T_π is a C^* algebra map; on the other hand, the range of T_π is dense in $C_\pi[V]$ since it contains the image of $C(V)$. Thus, T_π is actually an isomorphism of C^* algebras.

Theorem 1.2.3 *If $\pi \in \operatorname{Rep} V$ is nonzero then $\| \cdot \|_\infty = \| \cdot \|_\pi$ and the identity map on $C(V)$ extends continuously to a C^* algebra isomorphism $C[V] \to C_\pi[V]$.*   □

As we claimed, it follows in retrospect that we could have defined the C^* Clifford algebra of V to be the completion of $C(V)$ relative to the norm $\| \cdot \|_\pi$ arising from any nonzero $\pi \in \operatorname{Rep} V$ since all such norms are equal. In particular, we could have defined the C^* Clifford algebra of V to be the uniform closure of $\lambda(C(V))$ in $B(\mathbb{H}_\tau)$.

Just as the complex Clifford algebra $C(V)$ solves a universal mapping problem, so also does the C^* Clifford algebra $C[V]$. Recall that we suppress the canonical embedding $\phi : V \to C[V]$ and regard V as being included in $C[V]$.

Theorem 1.2.4 *If B is a unital C^* algebra and if $f : V \to B$ is a self-adjoint Clifford map, then there exists a unique (isometric) C^* algebra map $F : C[V] \to B$ such that $F \mid V = f$.*

Proof Firstly, Theorem 1.1.3 tells us that the unique algebra map $F : C(V) \to B$ such that $F \mid V = f$ is involution-preserving. Being plainly nonzero, this star-homomorphism F extends continuously to a C^* algebra isomorphism from $C[V]$ to $C_F[V] \subset B$ by virtue of Theorem 1.2.3. The resulting C^* algebra map $F : C[V] \to B$ is quite clearly unique given that its restriction to V should be f. Lastly, the isometric nature of $F : C[V] \to B$ is a consequence of its injectivity and can also be read from Theorem 1.2.3.   □

In order to formulate a standard functorial consequence of this universal mapping property, let V' be another real inner product space and let $g : V \to V'$ be an isometric linear map. Following g by the inclusion $V' \to C[V']$ produces a self-adjoint Clifford map $f : V \to C[V']$ which then extends to an isometric C^* algebra map F from $C[V]$ to $C[V']$ according to Theorem 1.2.4. We note that this C^* algebra map F extends the algebra map θ_g of Theorem 1.1.2 and continue the notation θ_g in place of F.

Theorem 1.2.5 *Each isometric linear map $g : V \to V'$ extends uniquely to an (isometric) C^* algebra map $\theta_g : C[V] \to C[V']$.* $\square$

In particular, if $g \in O(V)$ is an orthogonal transformation of V then the Bogoliubov automorphism θ_g of the complex Clifford algebra $C(V)$ extends to an automorphism θ_g of the C^* Clifford algebra $C[V]$ which we continue to speak of as a *Bogoliubov automorphism*. Thus we arrive at a faithful representation

$$\theta : O(V) \to \operatorname{Aut} C[V]$$

of the orthogonal group $O(V)$ by automorphisms of the C^* Clifford algebra $C[V]$.

As with the complex Clifford algebra, we set aside a separate symbol γ to denote the Bogoliubov automorphism θ_{-I} of the C^* Clifford algebra $C[V]$ induced by minus the identity $-I \in O(V)$. Since γ has period 2, we refer to it as the *grading automorphism* of $C[V]$: it engenders an eigendecomposition

$$C[V] = C^+[V] \oplus C^-[V]$$

in which the *even C^* Clifford algebra* $C^+[V]$ of V is fixed pointwise by γ and in which γ acts as minus the identity on the complementary subspace $C^-[V]$. Again, we speak of elements of $C^+[V]$ as being *even* and say that the elements of $C^-[V]$ are *odd*. The following elementary observation is worth noting.

Theorem 1.2.6 *$C^\pm[V]$ is the closure of $C^\pm(V)$ in $C[V]$.*

Proof The inclusion $C^\pm(V) \subset C^\pm[V]$ is plain. Let $a \in C^\pm[V]$ so that $\gamma(a) = \pm a$. Choose a sequence $(a_n : n > 0)$ in $C(V)$ so that $a_n \to a$ and note that $\gamma(a_n) \to \gamma(a)$ by norm continuity. The equations

$$a = \tfrac{1}{2}\big(a \pm \gamma(a)\big) = \lim \tfrac{1}{2}\big(a_n \pm \gamma(a_n)\big)$$

now express a as the limit of a sequence in $C^\pm(V)$. $\square$

Another consequence of the universal mapping property is that the C^* Clifford algebra $C[V]$ is insensitive to whether or not the real inner product space V is complete. Somewhat more explicitly, we have the following result.

Theorem 1.2.7 *If $\overline{V}$ is the Hilbert space completion of V then the inclusion $V \to \overline{V}$ extends to an isomorphism $C[V] \to C[\overline{V}]$ of C^* algebras.*

Proof First of all, the inclusion $V \to \overline{V}$ certainly induces a C^* algebra map $F : C[V] \to C[\overline{V}]$ by Theorem 1.2.5; observe that F is already injective. To see that F is also surjective, it is sufficient to construct a C^* algebra map $\Psi : C[\overline{V}] \to C[V]$ such that $F \circ \Psi = I$. For this, recall from Theorem 1.2.1 that the canonical inclusion $V \to C[V]$ is isometric and so extends continuously to define a self-adjoint Clifford map $\psi : \overline{V} \to C[V]$. According to the universal mapping property in Theorem 1.2.4, ψ extends to a C^* algebra map $\Psi : C[\overline{V}] \to C[V]$. The evident equality $F \circ \Psi = I$ on V extends to $\overline{V}$ by continuity and so to $C[\overline{V}]$ by uniqueness in the universal mapping property. $\qquad\qquad\qquad\square$

On the strength of this result, for many purposes it is harmless to suppose that V is a real Hilbert space. For this reason we shall often make this supposition in the sequel, though not without due notification of its being made.

With a view to drawing out further structural properties of the C^* Clifford algebra, we now return to the left regular representation λ of $C(V)$ on the Hilbert space completion $\mathbb{H}_\tau$ of $H_\tau = C(V)$ relative to the inner product $\langle \cdot \mid \cdot \rangle$ arising from the canonical trace τ. As with any nonzero star-representation of the complex Clifford algebra, λ automatically extends to an isometric representation

$$\lambda : C[V] \to B(\mathbb{H}_\tau)$$

of the C^* Clifford algebra, according to Theorem 1.2.3. The distinguished unit vector $\Omega = \mathbf{1}$ in $\mathbb{H}_\tau$ associates to λ a vector state, by sending $a \in C[V]$ to $\langle \lambda(a)\Omega \mid \Omega \rangle \in \mathbb{C}$. Note that if in fact $a \in C(V)$ then

$$\langle \lambda(a)\Omega \mid \Omega \rangle = \langle a \mid \mathbf{1} \rangle = \tau(\mathbf{1}^* a) = \tau(a)$$

so that on $C(V)$ this vector state agrees with the canonical trace. By continuity, this vector state is plainly the unique state of $C[V]$ with this property. Accordingly, we extend to it the notation τ: thus,

$$a \in C[V] \quad \Rightarrow \quad \tau(a) = \langle \lambda(a)\Omega \mid \Omega \rangle.$$

We call τ so defined the *trace* of the C^* Clifford algebra. It deserves this name by virtue of being central in the sense that if a and b lie in $C[V]$ then $\tau(ba) = \tau(ab)$. Indeed, if we choose sequences $(a_n : n > 0)$ and $(b_n : n > 0)$ in $C(V)$ such that $a_n \to a$ and $b_n \to b$ then continuity implies that the equality $\tau(b_n a_n) = \tau(a_n b_n)$ for τ on $C(V)$ becomes the equality $\tau(ba) = \tau(ab)$ for τ on $C[V]$ in the limit. Moreover, τ

continues to be even: by continuity, the equality $\tau \circ \gamma = \tau$ extends from $C(V)$ to $C[V]$ and the vanishing of τ on $C^-(V)$ implies its vanishing on $C^-[V]$ because of Theorem 1.2.6. Still more is true: any central state of $C[V]$ coincides with τ and is hence even. In fact, we have the following result.

Theorem 1.2.8 $C[V]$ *admits a unique even central state: its trace τ, of which any continuous central linear functional is a scalar multiple.*

Proof Suppose $\sigma : C[V] \to \mathbb{C}$ to be a continuous central linear functional. If $M \in \mathcal{F}^+(V)$ then the central linear functional $\sigma \mid C(M)$ is also even as in Theorem 1.1.12 and we may refashion the uniqueness argument of Theorem 1.1.7 to deduce that $\sigma = \sigma(\mathbf{1})\tau$ on $C(M)$. The remarks following Theorem 1.1.13 now imply that $\sigma = \sigma(\mathbf{1})\tau$ on $C(V)$. Finally, we deduce that $\sigma = \sigma(\mathbf{1})\tau$ on $C[V]$ by continuity. $\qquad\square$

Let us temporarily denote by Z_τ the set of all $a \in C[V]$ with the property that $\tau(ba) = 0$ whenever $b \in C[V]$: thus
$$Z_\tau = \big\{a \in C[V] : \tau\big(C[V] \cdot a\big) = 0\big\}.$$
Since τ is a central linear functional, $Z_\tau \subset C[V]$ is a bilateral ideal; since τ is continuous, $Z_\tau \subset C[V]$ is closed. The simplicity of $C[V]$ now intervenes: since $\mathbf{1} \notin Z_\tau$ it follows that $Z_\tau = 0$. This fact has important consequences, both directly for the trace τ and indirectly for the C^* Clifford algebra $C[V]$.

Regarding the trace τ, note that it satisfies a *Cauchy-Schwarz inequality* by virtue of being a state: explicitly, if $a, b \in C[V]$ then
$$|\tau(b^*a)|^2 \le \tau(a^*a)\tau(b^*b).$$
It follows from this and the vanishing of Z_τ that the trace τ is *faithful* in the sense that if $a \in C[V]$ satisfies $\tau(a^*a) = 0$ then $a = 0$. Indeed, if also $b \in C[V]$ then
$$|\tau(ba)|^2 \le \tau(a^*a)\tau(bb^*) = 0$$
whence $\tau(ba) = 0$ and so $a \in Z_\tau$.

Theorem 1.2.9 *The trace τ is faithful on $C[V]$.* $\qquad\square$

Regarding the C^* Clifford algebra, we contend that it is central in that its centre comprises precisely all scalar multiples of the identity. Although the analogous contention is valid for any simple unital C^* algebra, we offer a direct proof in this case. Let z lie in the centre of $C[V]$ and consider the continuous linear functional
$$\sigma : C[V] \to \mathbb{C} : a \mapsto \tau(az).$$

Since z lies in the centre of $C[V]$ and since τ is central, it follows that σ is central; in consequence, $\sigma = \sigma(1)\tau = \tau(z)\tau$ by Theorem 1.2.8. Now, if $b \in C[V]$ then

$$\tau\big(b(z - \tau(z)\mathbf{1})\big) = \tau(bz) - \tau(z)\tau(b)$$
$$= \sigma(b) - \tau(z)\tau(b) = 0$$

so that $z - \tau(z)\mathbf{1} \in Z_\tau = 0$ and $z = \tau(z)\mathbf{1}$ is a scalar multiple of the identity.

Theorem 1.2.10 *The C^* Clifford algebra $C[V]$ is central.* $\square$

Thus, the C^* Clifford algebra $C[V]$ is both central and simple. In fact, the even C^* Clifford algebra $C^+[V]$ inherits each of these properties. To see that this is so, we can mimic the arguments offered for the C^* Clifford algebra itself.

Theorem 1.2.11 *The even C^* Clifford algebra $C^+[V]$ is both central and simple.*

Proof For simplicity, we can proceed essentially after the pattern of Theorem 1.2.2 but using the following facts: that if $M \in \mathcal{F}^-(V)$ then $C^+(M)$ is simple as in the remarks at the close of the preceding section; that $C^+(V) = \bigcup\{C^+(M) : M \in \mathcal{F}^-(V)\}$ as in the remarks after Theorem 1.1.13. As before, centrality is a consequence of unital simplicity.
                                                                                   $\square$

However, the central simplicity of $C^+[V]$ can be established by other means; we turn now to examine a more productive method.

To begin, let $l \in V$ be a unit vector with linear span L and orthocomplementary hyperplane $L^\perp$. Recall from Theorem 1.1.18 that the self-adjoint Clifford map

$$\phi : L^\perp \rightarrow C^+(V) : v \mapsto ilv$$

gives rise by universality to an isomorphism of involutive algebras

$$\Phi : C(L^\perp) \rightarrow C^+(V)$$

given by $\Phi(a) = a$ if $a \in C^+(L^\perp)$ and $\Phi(a) = ila$ if $a \in C^-(L^\perp)$. Following this isomorphism with the inclusion of $C^+(V)$ in $C^+[V]$ yields a nonzero star-homomorphism $C(L^\perp) \rightarrow C^+[V]$ which we shall again denote by Φ for convenience. An application of either Theorem 1.2.3 or Theorem 1.2.4 now extends Φ to a C^* algebra map $C[L^\perp] \rightarrow C^+[V]$ which we shall continue to denote by Φ. This Φ is actually a C^* algebra isomorphism: it is injective since $C[L^\perp]$ is simple; it is surjective since its range contains $C^+(V)$ and is closed, Φ being a C^* algebra map. Note that continuity renders valid for $\Phi : C[L^\perp] \to C^+[V]$ the explicit formulae defining $\Phi : C(L^\perp) \to C^+(V)$. Thus, Theorem 1.1.18 for complex Clifford algebras has the following extension to their C^* algebra completions.

Theorem 1.2.12 *Let $l \in V$ be a unit vector and let L be its linear span. The map*

$$\Phi : C[L^\perp] \to C^+[V]$$

given by

$$a \in C^+[L^\perp] \quad \Rightarrow \quad \Phi(a) = a$$
$$a \in C^-[L^\perp] \quad \Rightarrow \quad \Phi(a) = \mathrm{i}la$$

is an isomorphism of C^* algebras. □

We see again that the even C^* Clifford algebra $C^+[V]$ is both central and simple. In addition, we deduce the following remarkable fact.

Theorem 1.2.13 *The C^* algebras $C[V]$ and $C^+[V]$ are isomorphic when V is infinite-dimensional.*

Proof Theorem 1.2.7 grants us the freedom to replace V by its completion; accordingly, we suppose V to be a real Hilbert space. Now, Theorem 1.2.12 tells us that if $l \in V$ is a unit vector and $L^\perp$ its orthocomplementary hyperplane, then the C^* algebras $C^+[V]$ and $C[L^\perp]$ are isomorphic. Next, the real Hilbert spaces $L^\perp$ and V are equidimensional and therefore isometrically isomorphic, whence Theorem 1.2.5 informs us that the C^* algebras $C[L^\perp]$ and $C[V]$ are isomorphic. The theorem follows at once. □

Of course, infinite-dimensionality of V is crucial to the validity of this result. Note that whereas the isomorphism in Theorem 1.2.12 is canonical once the unit vector l is chosen, that in Theorem 1.2.13 depends also on the choice of an isometric isomorphism between the completion of V and the corresponding orthocomplement of L.

The central nature of $C^+[V]$ implies that the grading automorphism γ of the C^* Clifford algebra $C[V]$ is not inner: the proof of this implication is essentially the same as that offered in Theorem 1.1.20 for the complex Clifford algebra. Thus: if $u \in C[V]$ is invertible and satisfies $\gamma(a) = uau^{-1}$ whenever $a \in C[V]$ then in fact u lies in the centre of $C^+[V]$ and is therefore a scalar, whence the absurdity that $-v = \gamma(v) = uvu^{-1} = v$ for all $v \in V$.

Theorem 1.2.14   *The grading automorphism γ of the C^* Clifford algebra $C[V]$ is not inner.*   □

In Section 4.2, we shall solve completely the general problem of determining which Bogoliubov automorphisms of the C^* Clifford algebra $C[V]$ are inner. Our solution to this problem requires an *auxiliary result* of independent interest, with which we close this section. Supposing V to be complete, this auxiliary result is as follows: if Z is any subspace of V then an even element of $C[V]$ commuting with each element of Z necessarily lies in $C^+[Z^\perp]$. Our discussion involves us in a largely algebraic detour, much of which could have been placed in the preceding section; it has been postponed until now to facilitate a coherent presentation.

First of all, we define the *adjoint representation*. If $u \in C(V)$ is invertible then Ad_u is the algebra automorphism of $C(V)$ given by

$$a \in C(V) \quad \Rightarrow \quad \mathrm{Ad}_u(a) = uau^{-1}.$$

We note that if $u \in C(V)$ is unitary then Ad_u is actually an automorphism of $C(V)$ as an involutive algebra. In like manner, we define the adjoint representation on $C[V]$ of its group of invertible elements, noting that its unitary elements are represented by automorphisms of $C[V]$ as a C^* algebra.

Now, recall from Theorem 1.1.18 that if $l \in V$ is a unit vector with linear span L then the map

$$\Phi : C(L^\perp) \rightarrow C^+(V)$$

given by

$$a \in C^+(L^\perp) \quad \Rightarrow \quad \Phi(a) = a$$
$$a \in C^-(L^\perp) \quad \Rightarrow \quad \Phi(a) = ila$$

is an isomorphism of involutive algebras. Let us consider the automorphism Ad_l as acting upon $C^+(V)$ by restriction. Having period 2, it splits $C^+(V)$ into the direct sum of its eigenspaces, with ± 1 as eigenvalues. In fact, we claim that these eigenspaces are

$$\ker(\mathrm{Ad}_l \mp I) = \Phi(C^\pm(L^\perp)).$$

For the justification of our claim, take $a \in C^+(V)$ and choose $M \in \mathcal{F}(V)$ so that $a \in C^+(M)$ by Theorem 1.1.13. Without loss, we may suppose that $v_1 = l \in M$ and shall then let M have $\{v_1, \ldots, v_m\}$ as an orthonormal basis. Put

$$a = \sum_{S \subset \mathbf{m}} \mu_S v_S$$

so that

$$lal = \sum_{S \subset \mathbf{m}} (-1)^{|S \cap 1|} \mu_S v_S$$

by the remarks made after Theorem 1.1.4. Noting that $(-1)^{|S \cap 1|}$ is -1 or $+1$ according to whether 1 does or does not lie in S, it follows that

$$lal = a \quad \Leftrightarrow \quad (\mu_S \neq 0 \quad \Rightarrow \quad 1 \notin S) \quad \Leftrightarrow \quad a \in C^+(L^\perp)$$

and

$$lal = -a \quad \Leftrightarrow \quad (\mu_S \neq 0 \quad \Rightarrow \quad 1 \in S) \quad \Leftrightarrow \quad a \in l \cdot C^-(L^\perp).$$

Recalling how Φ was defined, our claim is thus justified. We record it in the following form.

Theorem 1.2.15 *Let $l \in V$ be a unit vector with linear span L. If $a \in C^+(V)$ then*

$$lal = a \quad \Leftrightarrow \quad a \in C^+(L^\perp)$$
$$lal = -a \quad \Leftrightarrow \quad a \in l \cdot C^-(L^\perp).$$

$\square$

A precisely similar result holds true for the even C^* Clifford algebra: we claim that the eigenspaces of Ad_l on $C^+[V]$ are given by

$$\ker(\text{Ad}_l \mp I) = \Phi\big(C^\pm[L^\perp]\big)$$

where Φ is now the C^* algebra isomorphism of Theorem 1.2.12. We shall establish the identification of $\ker(\text{Ad}_l - I)$ as being $C^+[L^\perp] = \Phi\big(C^+[L^\perp]\big)$; the identification $\ker(\text{Ad}_l + I) = l \cdot C^-[L^\perp] = \Phi\big(C^-[L^\perp]\big)$ is quite similar. On the one hand, $\ker(\text{Ad}_l - I)$ contains $C^+(L^\perp)$ by Theorem 1.2.15 and is closed by continuity, so it contains $C^+[L^\perp]$. On the other hand, if $a \in C^+[V]$ is fixed by Ad_l and if $(a_n : n > 0)$ is a sequence in $C^+(V)$ such that $a_n \to a$ then also $l\,a_n l \to lal = a$ by continuity, whence $a = \lim \frac{1}{2}(a_n + l\,a_n l)$; since $a_n + l\,a_n l$ lies in $C^+(L^\perp)$ by Theorem 1.2.15, it follows that a lies in $C^+[L^\perp]$.

Theorem 1.2.16 *Let $l \in V$ be a unit vector with linear span L. If $a \in C^+[V]$ then*

$$lal = a \quad \Leftrightarrow \quad a \in C^+[L^\perp]$$
$$lal = -a \quad \Leftrightarrow \quad a \in l \cdot C^-[L^\perp].$$

$\square$

We remark that each of these two results involving the unit vector $l \in V$ with orthocomplement $L^\perp$ admits an instructive reformulation. For the complex Clifford algebra: an even element of $C(V)$ lies in $C^+(L^\perp)$ iff it commutes with l; an odd element of $C(V)$ lies in $C^-(L^\perp)$ iff it anticommutes with l. For the C^* Clifford algebra likewise: an even element of $C[V]$ lies in $C^+[L^\perp]$ iff it commutes with l; an odd element of $C[V]$ lies in $C^-[L^\perp]$ iff it anticommutes with l. We remark in addition that these two results also follow easily from Theorem 1.1.18 and Theorem 1.2.12.

The next phase of our essentially algebraic detour is to associate to any finite-dimensional subspace M of V a *conditional expectation*

$$p = p_M : C^+(V) \to C^+(M^\perp)$$

by which we mean a surjective linear map preserving multiplicative identities and having the property that

$$p(bac) = b\,p(a)c$$

whenever $a \in C^+(V)$ and $b, c \in C^+(M^\perp)$. In particular, note that p restricts to $C^+(M^\perp)$ as the identity map, since if $b \in C^+(M^\perp)$ then

$$p(b) = p(b\mathbf{1}) = b\,p(\mathbf{1}) = b.$$

Note also that $C^+(M^\perp)$ comprises precisely all elements of $C^+(V)$ that commute with each element of M; this is readily seen by arguing as for Theorem 1.2.15, letting l there run through an orthonormal basis for M.

To begin our construction, let M have $\{v_1, \ldots, v_m\}$ as an orthonormal basis. For $j \in \mathbf{m}$ let

$$p_j : C^+(V) \to C^+(v_j^\perp)$$

be the eigenprojection with eigenvalue $+1$ for $\mathrm{Ad}\,_{v_j}$: explicitly,

$$a \in C^+(V) \quad \Rightarrow \quad p_j(a) = \tfrac{1}{2}(a + v_j a\,v_j).$$

The Clifford relations ensure that the idempotents $p_1, \ldots, p_m$ commute, whence their product $p_M := p_1 \ldots p_m$ is also an idempotent. In light of the fact that $v_1^\perp \cap \ldots \cap v_m^\perp = M^\perp$ it is plain that p_M actually maps $C^+(V)$ onto $C^+(M^\perp)$. Of course, p_M is identity-preserving: in fact, if $a \in C^+(V)$ commutes with each element of M then $p_j(a) = a$ for all $j \in \mathbf{m}$ and so $p_M(a) = a$. Now let a lie in $C^+(V)$ and let $b, c \in C^+(M^\perp)$: if $j \in \mathbf{m}$ then

$$
\begin{aligned}
2p_j(bac) &= bac + v_j bac\,v_j \\
&= bac + bv_j a\,v_j c \\
&= b(a + v_j a\,v_j)c \\
&= 2b\,p_j(a)c
\end{aligned}
$$

since $v_j \in M$ commutes with elements of $C^+(M^\perp)$ by virtue of the Clifford relations; consequently,

$$p_M(bac) = b\,p_M(a)c.$$

Thus, we have indeed constructed a conditional expectation p_M from $C^+(V)$ onto $C^+(M^\perp)$. Of course, a question at once presents itself: namely, whether p_M is independent of the orthonormal basis $\{v_1, \ldots, v_m\}$

from which it was constructed. This natural question has an affirmative answer: in fact, p_M is precisely orthogonal projection of $C^+(V)$ on $C^+(M^\perp)$ relative to the inner product $\langle \cdot \mid \cdot \rangle$ defined by the trace τ as in Theorem 1.1.17. To see this, note first that if $j \in \mathbf{m}$ then p_j is self-adjoint: if $\xi, \eta \in C^+(V)$ then

$$
\begin{aligned}
2\langle p_j(\xi) \mid \eta \rangle &= 2\tau(\eta^* p_j(\xi)) \\
&= \tau(\eta^*\xi) + \tau(\eta^* v_j \xi v_j) \\
&= \tau(\eta^*\xi) + \tau(v_j \eta^* v_j \xi) \\
&= 2\tau(p_j(\eta)^*\xi) \\
&= 2\langle \xi \mid p_j(\eta) \rangle
\end{aligned}
$$

since τ is central. Since $p_1, \ldots, p_m$ commute, the idempotent $p_M = p_1 \ldots p_m$ is also self-adjoint. This is enough to support our affirmative answer; if $\zeta \in C^+(V)$ then $\zeta - p_M(\zeta)$ is orthogonal to $p_M(\zeta) \in C^+(M^\perp)$. We summarize this particular phase of our detour as follows.

Theorem 1.2.17 *If $M \in \mathcal{F}(V)$ then a conditional expectation p_M from $C^+(V)$ onto $C^+(M^\perp)$ is given by orthogonal projection relative to the inner product arising from the trace.* $\square$

Again, these findings have counterparts applicable to the even C^* Clifford algebra. In fact, note from Theorem 1.2.1 that if $a \in C^+(V)$ then $\|p_j(a)\|_\infty \le \|a\|_\infty$ for each $j \in \mathbf{m}$ and so $\|p_M(a)\|_\infty \le \|a\|_\infty$. As a consequence, the maps $p_1, \ldots, p_m$ and p_M extend continuously to $C^+[V]$ in such a way that the extensions of $p_1, \ldots, p_m$ are given by the same formulae as before and their product is the extension

$$
p_M : C^+[V] \to C^+[M^\perp].
$$

Of course, p_M continues to be an identity-preserving linear map. It also satisfies

$$
p_M(bac) = b\, p_M(a) c
$$

whenever $a \in C^+[V]$ and $b, c \in C^+[M^\perp]$; this follows by continuity. We have thus established the following.

Theorem 1.2.18 *If $M \in \mathcal{F}(V)$ then the continuous extension of p_M is a conditional expectation from $C^+[V]$ onto $C^+[M^\perp]$.* $\square$

We must remark on one special property of these conditional expectations, taking first the even complex Clifford algebras. Suppose W to be a (not necessarily finite-dimensional) subspace of V containing the finite-dimensional subspace M. The decomposition $p_M = p_1 \ldots p_m$ applies equally to each of $C^+(V) \to C^+(M^\perp)$ and $C^+(W) \to C^+(W \cap M^\perp)$; accordingly, the former restricts to the latter. Likewise, the conditional expectation $C^+[V] \to C^+[M^\perp]$ restricts to the conditional expectation $C^+[W] \to C^+[W \cap M^\perp]$. In particular, we have the following result, recorded for convenience.

Theorem 1.2.19 *Let $p_M : C^+[V] \to C^+[M^\perp]$ be the conditional expectation associated to $M \in \mathcal{F}(V)$. If W is a subspace of V containing M then $p_M\left(C^+[W]\right) \subset C^+[W \cap M^\perp]$.* □

At last, we are able to present a proof of the promised *auxiliary result*. Recall that this result supposes V to be complete; it asserts that if $a \in C^+[V]$ commutes with each element of a subspace $Z \subset V$ then in fact $a \in C^+[Z^\perp]$. Observe at once that we may assume Z to be closed, since Z and its closure share the same orthogonal space and since elements of $C^+[V]$ commuting with Z commute also with its closure. Now, choose a sequence $(a_n : n > 0)$ in $C^+(V)$ such that $a_n \to a$ and for each $n > 0$ choose $M_n \in \mathcal{F}(V)$ such that $a_n \in C^+(M_n)$. Let M'_n be the orthogonal projection of M_n in Z and put $M''_n = M'_n + M_n$. Of course, M''_n is a finite-dimensional subspace of V; additionally, the orthocomplement $M''_n \ominus M'_n = M''_n \cap (M'_n)^\perp$ of M'_n in M''_n is contained in $Z^\perp$. We now make use of the conditional expectation $p_{M'_n}$ from $C^+[V]$ onto $C^+[(M'_n)^\perp]$ as in Theorem 1.2.18. Since $a \in C^+[V]$ commutes with each element of $Z \supset M'_n$ it follows from the construction of $p_{M'_n}$ that in fact $p_{M'_n}(a) = a$. Since $a_n \in C^+(M_n) \subset C^+(M''_n)$ it follows from Theorem 1.2.19 that in fact $p_{M'_n}(a_n) \in C^+(M''_n \ominus M'_n) \subset C^+[Z^\perp]$. The norm-decreasing nature of $p_{M'_n}$ now implies that

$$\|p_{M'_n}(a_n) - a\|_\infty = \|p_{M'_n}(a_n - a)\|_\infty \leq \|a_n - a\|_\infty$$

and a is the limit of a sequence in $C^+[Z^\perp]$. Of course, this forces a itself to lie in $C^+[Z^\perp]$ as asserted.

Theorem 1.2.20 *If V is a real Hilbert space of which Z is a subspace, then elements of $C^+[V]$ commuting with Z necessarily lie in $C^+[Z^\perp]$.* □

The converse of this result is quite elementary: the elements of any subspace $Z \subset V$ commute with every element of $C^+[Z^\perp]$ as follows by continuity from the Clifford relations.

We have referred to Theorem 1.2.20 as an auxiliary result and so it is, for us: we shall make use of it in Section 4.2, in confronting the possibility that the real Hilbert space V may not be separable. Nevertheless, both the result itself and the detour leading up to it clearly have independent merit.

1.3 vN Clifford algebras

Naturally associated to the real inner product space V there is yet another Clifford algebra, still larger than its C^* Clifford algebra. Recall that the complex Clifford algebra $C(V)$ acts by the left regular representation λ on its completion $\mathbb{H}_\tau$ relative to the inner product arising from its canonical trace τ. The von Neumann algebra $\mathcal{A}[V]$ generated by the range of λ in $B(\mathbb{H}_\tau)$ is called the vN Clifford algebra of V and forms the principal object of study in this section. Of course, if V is finite-dimensional then the range of λ is closed in each of the standard operator topologies, whence $\mathcal{A}[V]$ and $C(V)$ may be identified. Accordingly, throughout the whole of this section, we shall suppose V to be *infinite-dimensional*.

A little more explicitly, recall that $\mathbb{H}_\tau$ denotes the Hilbert space completion of $H_\tau = C(V)$ equipped with the inner product defined by

$$\xi, \eta \in C(V) \quad \Rightarrow \quad \langle \xi \mid \eta \rangle = \tau(\eta^* \xi)$$

and that the left regular representation λ of $C[V] \supset C(V)$ on $\mathbb{H}_\tau$ is determined by the prescription

$$a \in C(V), \ \zeta \in H_\tau \quad \Rightarrow \quad \lambda(a)\zeta = a\zeta.$$

As a matter of definition, we write

$$\mathcal{A}[V] := \lambda\big(C(V)\big)'' \subset B(\mathbb{H}_\tau)$$

for the bicommutant of the image of $C(V)$ in $B(\mathbb{H}_\tau)$ under λ. Thus, $\mathcal{A}[V]$ comprises precisely all bounded linear operators on $B(\mathbb{H}_\tau)$ that commute with each operator commuting with $\lambda(a)$ whenever $a \in C(V)$. According to the von Neumann bicommutant theorem, $\mathcal{A}[V]$ is precisely the closure of $\lambda\big(C(V)\big)$ in $B(\mathbb{H}_\tau)$ relative to each of the standard operator topologies other than that defined by the operator norm. Being a von Neumann algebra, $\mathcal{A}[V]$ will be called the *vN Clifford algebra* of V. Of course, $\mathcal{A}[V]$ contains the image $C_\lambda[V]$ of the C^* Clifford algebra $C[V]$ under λ; indeed, $\mathcal{A}[V]$ is precisely $C_\lambda[V]''$.

As was true of the C^* Clifford algebra, the vN Clifford algebra is insensitive to the completeness or otherwise of the underlying real inner product space; we proceed to establish this convenient fact at once.

It will clarify matters if we reconsider the Hilbert space $\mathbb{H}_\tau$ and denote it by $\mathbb{H}(V)$ temporarily. The prescription

$$\xi, \eta \in C[V] \quad \Rightarrow \quad \langle \xi \mid \eta \rangle = \tau(\eta^* \xi)$$

defines an inner product on the C^* Clifford algebra $C[V]$ since its trace τ is faithful by Theorem 1.2.9; we temporarily write $\mathbb{H}[V]$ for the resulting Hilbert space completion of the C^* Clifford algebra.

Theorem 1.3.1   *The inclusion of $C(V)$ in $C[V]$ extends by continuity to an isometric isomorphism between $\mathbb{H}(V)$ and $\mathbb{H}[V]$.*

Proof   It is clearly enough to show that $C(V) \subset \mathbb{H}[V]$ is dense. Note first that on $C[V]$ the operator norm $\| \cdot \|_\infty$ dominates the norm $\| \cdot \|_\tau$ arising from $\langle \cdot \mid \cdot \rangle$: indeed, if $a \in C[V]$ then
$$\|a\|_\tau^2 = \tau(a^*a) \le \|a^*a\|_\infty = \|a\|_\infty^2.$$
Now, let $\zeta \in \mathbb{H}[V]$ and let $\epsilon > 0$. There exists $\eta \in C[V]$ such that $\|\zeta - \eta\|_\tau \le \frac{1}{2}\epsilon$ and then there exists $\xi \in C(V)$ such that $\|\eta - \xi\|_\infty \le \frac{1}{2}\epsilon$ whence $\|\eta - \xi\|_\tau \le \frac{1}{2}\epsilon$ from above. Finally, $\|\zeta - \xi\|_\tau \le \epsilon$ by the triangle inequality. $\square$

Of course, it goes without saying that this canonical isometric isomorphism $\mathbb{H}(V) \to \mathbb{H}[V]$ intertwines the left regular representations of $C(V)$ and $C[V]$. This being the case, either $\mathbb{H}(V)$ or $\mathbb{H}[V]$ may be taken as $\mathbb{H}_\tau$ and either may be taken to carry the vN Clifford algebra of V.

From Theorem 1.2.7 we recall also that the inclusion $C(V) \subset C(\overline{V})$ extends continuously to a canonical isomorphism $F : C[V] \to C[\overline{V}]$. Theorem 1.2.8 on the uniqueness of τ as a central state implies that F pulls back the trace on $C[\overline{V}]$ to the trace on $C[V]$. As a result, F is isometric and so extends by continuity to a canonical isometric isomorphism $U : \mathbb{H}[V] \to \mathbb{H}[\overline{V}]$. Plainly, this isomorphism intertwines the left regular representations of $C[V]$ and $C[\overline{V}]$ identified by F: more explicitly, if $a \in C[V]$ then $U\lambda(a) = \lambda(Fa)U$. It follows that the isomorphism
$$B(\mathbb{H}[V]) \to B(\mathbb{H}[\overline{V}]) : A \mapsto UAU^*$$
sends $C_\lambda[V]$ to $C_\lambda[\overline{V}]$ and hence sends $\mathcal{A}[V] = C_\lambda[V]''$ to $\mathcal{A}[\overline{V}] = C_\lambda[\overline{V}]''$. Identifying a complex Clifford algebra with its image in the corresponding vN Clifford algebra, the outcome of these deliberations may be formulated in the following manner.

Theorem 1.3.2   *If $\overline{V}$ is the Hilbert space completion of V then the inclusion $C(V) \subset C(\overline{V})$ extends to a canonical isomorphism $\mathcal{A}[V] \to \mathcal{A}[\overline{V}]$ of von Neumann algebras.* $\square$

Again, this result gives us the right of supposing V to be a real Hilbert space for many purposes; as before, this right will not be invoked without due warning.

Now, it turns out that $\mathcal{A}[V]$ is not merely a von Neumann algebra: it is actually a *factor* in the sense that its centre comprises precisely all scalar operators; in fact, $\mathcal{A}[V]$ is a factor of type II_1 in terms of the classification due to Murray and von Neumann. These facts are conveniently derived by studying the extension of τ to a unique normal trace on the vN Clifford algebra.

With this aim in view, consider the vector state associated to the von Neumann algebra $\mathcal{A}[V] \subset B(\mathbb{H}_\tau)$ by the standard unit vector $\Omega = 1 \in \mathbb{H}_\tau$ assigning to $A \in \mathcal{A}[V]$ the scalar $\langle A \cdot \Omega \mid \Omega \rangle$. This state restricts to $C_\lambda[V] \equiv C[V]$ as the canonical trace and will therefore be denoted by the same symbol τ. Thus, τ is given by the rule

$$\tau : \mathcal{A}[V] \rightarrow \mathbb{C} : A \mapsto \langle A\Omega \mid \Omega \rangle.$$

Being a vector state, τ is of course weakly continuous, thus ultraweakly continuous and so normal. Either version of continuity renders τ unique subject to its agreeing with the canonical trace when restricted to $C(V)$.

We claim that τ is *central* on $\mathcal{A}[V]$ in the sense that $\tau(BA) = \tau(AB)$ whenever $A, B \in \mathcal{A}[V]$. To see this, choose a net $(a_j : j \in \mathcal{J})$ in $C(V)$ such that $\lambda(a_j) \overset{w}{\rightarrow} A$ and let $b \in C(V)$. As multiplication on either side is continuous in the weak operator topology, we find that $\lambda(b)\lambda(a_j) \overset{w}{\rightarrow} \lambda(b)A$ and $\lambda(a_j)\lambda(b) \overset{w}{\rightarrow} A\lambda(b)$. The facts that τ is weakly continuous on $\mathcal{A}[V]$ and central on $C(V)$ now imply that

$$
\begin{aligned}
\tau\big(\lambda(b)A\big) &= \lim_j \tau\big(\lambda(b)\lambda(a_j)\big) \\
&= \lim_j \tau(b\,a_j) \\
&= \lim_j \tau(a_j b) \\
&= \lim_j \tau\big(\lambda(a_j)\lambda(b)\big) \\
&= \tau\big(A\lambda(b)\big).
\end{aligned}
$$

A similar repetition, replacing b by a net $(b_j : j \in \mathcal{J})$ in $C(V)$ such that $\lambda(b_j) \overset{w}{\rightarrow} B$, shows that $\tau(BA) = \tau(AB)$ as claimed.

As a central normal state on the vN Clifford algebra, τ is unique; in fact, the following is true.

Theorem 1.3.3 $\mathcal{A}[V]$ *has a unique central normal state: its trace* τ, *of which any ultraweakly continuous central linear functional is a scalar multiple.*

Proof Let $\sigma : \mathcal{A}[V] \rightarrow \mathbb{C}$ be an ultraweakly continuous central linear functional. As in Theorem 1.2.8 we see that the equality $\sigma = \sigma(\mathbf{1})\tau$ holds on $C(V)$; this equality continues to hold on the whole of $\mathcal{A}[V]$ by ultraweak continuity. $\square$

As noted in the theorem, we shall refer to $\tau : \mathcal{A}[V] \to \mathbb{C}$ as the *trace* on the vN Clifford algebra.

Observe that the standard unit vector $\Omega \in \mathbb{H}_\tau$ is *cyclic* for $\mathcal{A}[V]$ in the sense that $\{A \cdot \Omega : A \in \mathcal{A}[V]\}$ is dense in $\mathbb{H}_\tau$: in fact, $\{\lambda(a)\Omega : a \in C(V)\} = C(V)$ is already dense in $\mathbb{H}_\tau$. Since τ is central, Ω is also a *trace vector* for $\mathcal{A}[V]$ in the sense that if $A, B \in \mathcal{A}[V]$ then

$$\langle BA\Omega \mid \Omega \rangle = \langle AB\Omega \mid \Omega \rangle$$

since here the left side is $\tau(BA)$ and the right side is $\tau(AB)$. As with any cyclic trace vector, it follows that Ω is *separating* for $\mathcal{A}[V]$ in the sense that if $Z \in \mathcal{A}[V]$ and $Z\Omega = 0$ then $Z = 0$. To see that this is so, note that if $A \in \mathcal{A}[V]$ then

$$\begin{aligned}
\|ZA\Omega\|^2 &= \langle ZA\Omega \mid ZA\Omega \rangle \\
&= \langle A^*Z^*ZA\Omega \mid \Omega \rangle \\
&= \langle AA^*Z^*Z\Omega \mid \Omega \rangle \\
&= 0
\end{aligned}$$

since Ω is a trace vector, so $ZA\Omega = 0$ and therefore $Z = 0$ since Ω is cyclic.

Theorem 1.3.4   *The standard unit vector $\Omega \in \mathbb{H}_\tau$ is a cyclic trace (hence separating) vector for $\mathcal{A}[V]$.*   □

We note that the separating nature of Ω for $\mathcal{A}[V]$ amounts to the faithfulness of τ on $\mathcal{A}[V]$. Indeed, if $Z \in \mathcal{A}[V]$ then

$$\|Z\Omega\|^2 = \langle Z\Omega \mid \Omega Z \rangle = \langle Z^*Z\Omega \mid \Omega \rangle = \tau(Z^*Z)$$

so that if $\tau(Z^*Z) = 0$ then $Z\Omega = 0$ and therefore $Z = 0$.

Theorem 1.3.5   *The trace τ is faithful on $\mathcal{A}[V]$.*   □

We are now able to show that the von Neumann algebra $\mathcal{A}[V]$ has scalar centre and is hence a *factor*. Let Z lie in the centre of $\mathcal{A}[V]$ and consider the linear functional

$$\sigma : \mathcal{A}[V] \to \mathbb{C} : A \mapsto \tau(AZ).$$

Since multiplication on either side is continuous in the weak operator topology, σ is weakly continuous and hence certainly normal. Since Z lies in the centre of the vN Clifford algebra, τ being central forces σ to be central. Theorem 1.3.3 now tells us that $\sigma = \sigma(\mathbf{1})\tau = \tau(Z)\tau$. The Hermitian property

$$\tau(Z^*) = \langle Z^*\Omega \mid \Omega \rangle = \langle \Omega \mid Z\Omega \rangle = \overline{\tau(Z)}$$

implies in particular that

$$\tau(Z^*Z) = \sigma(Z^*) = \tau(Z)\tau(Z^*) = |\tau(Z)|^2.$$

Consequently,

$$\tau\big((Z - \tau(Z)\mathbf{1})^*(Z - \tau(Z)\mathbf{1}) \big)$$
$$= \tau\big(Z^*Z - \overline{\tau(Z)}Z - \tau(Z)Z^* + |\tau(Z)|^2\mathbf{1} \big)$$
$$= 0$$

and therefore $Z = \tau(Z)\mathbf{1}$ by Theorem 1.3.5.

Theorem 1.3.6 $\mathcal{A}[V]$ *is a factor.* $\qquad\square$

The trace also allows us to show that $\mathcal{A}[V]$ is of type II_1 in the classification of factors due to Murray and von Neumann. Recall that this classification is based upon the dimensional equivalence of projections: two projections E and F in a von Neumann algebra are said to be (*dimensionally*) *equivalent* if and only if the von Neumann algebra contains an element G such that $GG^* = E$ and $G^*G = F$; the von Neumann algebra is said to be *finite* if and only if the only projection equivalent to $\mathbf{1}$ is $\mathbf{1}$ itself. Recall also that a factor has type II_1 if and only if it is both infinite-dimensional and finite.

Theorem 1.3.7 $\mathcal{A}[V]$ *has type* II_1.

Proof As $\mathcal{A}[V]$ is infinite-dimensional, we need only show that it is finite. Thus, let $E \in \mathcal{A}[V]$ be a projection and let $G \in \mathcal{A}[V]$ be such that $GG^* = E$ and $G^*G = \mathbf{1}$. Since τ is central it annihilates the difference $G^*G - GG^*$. This difference equals $\mathbf{1} - E$ and is therefore positive, of the form A^*A for some $A \in \mathcal{A}[V]$. The faithfulness of τ in Theorem 1.3.5 now implies that $A = 0$ and so $E = \mathbf{1}$. $\qquad\square$

Moving on, we next consider Bogoliubov automorphisms of the vN Clifford algebra. Actually, it is rather convenient to construct them somewhat indirectly. Firstly, we show that if $g \in O(V)$ is an orthogonal transformation then the Bogoliubov automorphism θ_g of $C(V)$ extends to a unitary operator U_g on $\mathbb{H}_\tau$. Secondly, we show that the assignment to $A \in \mathcal{A}[V]$ of $U_g A U_g^* \in B(\mathbb{H}_\tau)$ actually defines an automorphism of $\mathcal{A}[V]$ and may be taken as the induced Bogoliubov automorphism of the vN Clifford algebra. Incidentally, it is worth remarking that this is part of a general picture: if $\mathbb{H}$ is a Hilbert space and if $\mathcal{A} \subset B(\mathbb{H})$ is a von Neumann algebra with a separating cyclic vector, then each automorphism Φ of $\mathcal{A}$ is *spatial* in that there exists a unitary operator U on $\mathbb{H}$ with the property that $\Phi(A) = UAU^*$ whenever $A \in \mathcal{A}$.

We begin by letting $g \in O(V)$ be an orthogonal transformation of V. Recall that g extends to define an automorphism θ_g of the complex Clifford algebra $C(V)$ preserving its main involution. Since θ_g is an algebra automorphism, $\tau \circ \theta_g$ is a normalized central linear functional on $C(V)$ and hence equals τ itself. Since θ_g also preserves the main involution, if $\zeta \in H_\tau$ then

$$\|\theta_g\zeta\|^2 = \tau\big((\theta_g\zeta)^*(\theta_g\zeta)\big)$$
$$= \tau\big(\theta_g(\zeta^*\zeta)\big)$$
$$= \tau(\zeta^*\zeta)$$
$$= \|\zeta\|^2$$

as a consequence of which θ_g extends by continuity to a unitary operator U_g on $\mathbb{H}_\tau$.

We continue by claiming that the unitary operator U_g on $\mathbb{H}_\tau$ implements the Bogoliubov automorphism θ_g of $C(V)$ in the left regular representation λ: indeed, if $a \in C(V)$ and $\zeta \in H_\tau$ then

$$\lambda(\theta_g a) \circ U_g(\zeta) = (\theta_g a)(\theta_g \zeta)$$
$$= \theta_g(a\zeta)$$
$$= U_g(a\zeta)$$
$$= U_g \circ \lambda(a)\zeta$$

so that by continuity

$$\lambda(\theta_g a) = U_g \lambda(a) U_g^*$$

on $\mathbb{H}_\tau$ as claimed. Let us briefly summarize our progress thus far in the form of a theorem.

Theorem 1.3.8 *If $g \in O(V)$ then the Bogoliubov automorphism θ_g of $C(V)$ is implemented by its unitary extension U_g on $\mathbb{H}_\tau$ in the left regular representation λ.* $\square$

Conjugation by U_g now defines an (inner) automorphism of $B(\mathbb{H}_\tau)$. We claim that this map

$$B(\mathbb{H}_\tau) \to B(\mathbb{H}_\tau) : A \mapsto U_g A U_g^*$$

restricts to an automorphism of the von Neumann algebra $\mathcal{A}[V]$. Plainly, it is enough to show that if A lies in $\mathcal{A}[V]$ then so does $U_g A U_g^*$. For this, let $(a_j : j \in \mathcal{J})$ be a net in $C(V)$ such that $\lambda(a_j) \overset{w}{\to} A$. On the one hand, $U_g \lambda(a_j) U_g^* = \lambda(\theta_g a_j)$ lies in $C(V)$ by Theorem 1.3.8; on the other hand, $U_g \lambda(a_j) U_g^* \overset{w}{\to} U_g A U_g^*$ by weak continuity of multiplication when all variables but one are fixed. The operator $U_g A U_g^*$ consequently lies in the weak closure $\mathcal{A}[V]$ of $\lambda(C(V)) \subset B(\mathbb{H}_\tau)$ as needed to support our claim. Of course, the automorphism of $\mathcal{A}[V]$ sending A to $U_g A U_g^*$ is unique subject to the condition that its restriction to the weakly dense subalgebra $\lambda(C(V)) \equiv C(V)$ should coincide with θ_g. Accordingly, we refer to it as the *Bogoliubov automorphism* θ_g of $\mathcal{A}[V]$: explicitly, it is defined by the rule

$$A \in \mathcal{A}[V] \quad \Rightarrow \quad \theta_g(A) = U_g A U_g^*.$$

In common with automorphisms of von Neumann algebras generally, θ_g is ultraweakly continuous; being spatial, it is also weakly continuous. As for the C^* Clifford algebra and the complex Clifford algebra before it, assigning Bogoliubov automorphisms of the vN Clifford algebra $\mathcal{A}[V]$ to orthogonal transformations of V yields a group homomorphism

$$\theta : O(V) \to \operatorname{Aut} \mathcal{A}[V].$$

As before, we single out for special mention the automorphisms induced by minus the identity: we write Γ in place of $U_{-I} \in \operatorname{Aut} \mathbb{H}_\tau$ and write γ in place of $\theta_{-I} \in \operatorname{Aut} \mathcal{A}[V]$. Note that Γ is a *symmetry* on $\mathbb{H}_\tau$: it satisfies $\Gamma^* = \Gamma = \Gamma^{-1}$. Note also that Γ implements γ in the left regular representation:

$$A \in \mathcal{A}[V] \quad \Rightarrow \quad \gamma(A) = \Gamma A \Gamma.$$

The *grading automorphism* γ of $\mathcal{A}[V]$ induces an eigendecomposition

$$\mathcal{A}[V] = \mathcal{A}^+[V] \oplus \mathcal{A}^-[V]$$

in which γ acts as the identity on the *even vN Clifford algebra* $\mathcal{A}^+[V]$ and as minus the identity on the complementary subspace $\mathcal{A}^-[V]$. Likewise, the *grading operator* Γ on $\mathbb{H}_\tau$ induces an orthogonal eigendecomposition

$$\mathbb{H}_\tau = \mathbb{H}_\tau^+ \oplus \mathbb{H}_\tau^-$$

in which $\mathbb{H}_\tau^\pm = (I \pm \Gamma)\mathbb{H}_\tau$ is of course the closure of $H_\tau^\pm = C^\pm(V)$ in $\mathbb{H}_\tau$. As usual, we refer to elements of $\mathcal{A}^+[V]$ and $\mathbb{H}_\tau^+$ as being *even* and to elements of $\mathcal{A}^-[V]$ and $\mathbb{H}_\tau^-$ as being *odd*.

The von Neumann algebra $\mathcal{A}^+[V]$ merits a little closer scrutiny. It so happens that $\mathcal{A}^+[V]$ is actually isomorphic to $\mathcal{A}[V]$ itself and is therefore also a type II_1 factor, as a result of which the grading automorphism γ of $\mathcal{A}[V]$ is not inner. As a first step in the direction of establishing these facts, we consider alternative descriptions of the even vN Clifford algebra.

First, we claim that $\mathcal{A}^+[V]$ is precisely the von Neumann algebra on $\mathbb{H}_\tau$ generated by the even complex Clifford algebra $C^+(V)$ in the representation λ: thus,

$$\mathcal{A}^+[V] = \lambda(C^+(V))'' \subset B(\mathbb{H}_\tau).$$

The inclusion $C^+(V) \subset C(V)$ immediately implies of course that $\lambda(C^+(V))'' \subset \lambda(C(V))''$. For the reverse inclusion, let $A \in \mathcal{A}^+[V]$ and choose a net $(a_j : j \in \mathcal{J})$ in $C(V)$ such that $\lambda(a_j) \xrightarrow{w} A$. Upon averaging with

$$\lambda(\gamma a_j) = \Gamma\lambda(a_j)\Gamma \xrightarrow{w} \Gamma A \Gamma = \gamma(A) = A$$

we may suppose that $\gamma(a_j) = a_j$ for each $j \in \mathcal{J}$. Thus, A lies in the weak closure $\lambda(C^+(V))''$ and we are done.

Theorem 1.3.9 $\mathcal{A}^+[V] = \lambda(C^+(V))'' \subset B(\mathbb{H}_\tau)$. $\qquad\qquad\qquad\square$

Second, we claim that $\mathcal{A}^+[V]$ may be naturally identified with a von Neumann algebra on $\mathbb{H}_\tau^+$. In order to formulate precisely which von Neumann algebra and how it is to be identified, note that the even and odd spaces $\mathbb{H}_\tau^\pm \subset \mathbb{H}_\tau$ are invariant under $C^+(V)$ in the left regular representation λ since if $a \in C^+(V)$ then

$$\lambda(a) = \lambda(\gamma a) = \Gamma\lambda(a)\Gamma.$$

Let us agree to write

$$\lambda^+ : C^+(V) \to B(\mathbb{H}_\tau^+)$$
$$\lambda^- : C^+(V) \to B(\mathbb{H}_\tau^-)$$

for the corresponding representations of the even complex Clifford algebra. Our contention now is the following.

Theorem 1.3.10 *Restriction from $\mathbb{H}_\tau$ to $\mathbb{H}_\tau^+$ defines an isomorphism from $\mathcal{A}^+[V]$ to $\lambda^+(C^+(V))'' \subset B(\mathbb{H}_\tau^+)$.*

Proof If $A \in \mathcal{A}^+[V]$ then $T(A) = A \mid \mathbb{H}_\tau^+$ maps $\mathbb{H}_\tau^+$ to itself. As a consequence, we obtain a star-representation $T : \mathcal{A}^+[V] \to B(\mathbb{H}_\tau^+)$. Note that T is faithful, since $\Omega \in \mathbb{H}_\tau^+$ is separating by Theorem 1.3.4. Note also that T is weakly continuous and hence normal, so that its range $\operatorname{ran} T \subset B(\mathbb{H}_\tau^+)$ is a von Neumann algebra. In fact, $\operatorname{ran} T = \lambda^+(C^+(V))''$: on the one hand, $\operatorname{ran} T \supset \lambda^+(C^+(V))$ since $\mathcal{A}^+[V] \supset \lambda(C^+(V))$; on the other, the usual averaging argument yields $\operatorname{ran} T \subset \lambda^+(C^+(V))''$. This concludes the proof.

$\qquad\qquad\qquad\qquad\qquad\qquad\qquad\qquad\qquad\qquad\qquad\qquad\square$

Now, the even vN Clifford algebra $\mathcal{A}^+[V]$ is in fact isomorphic to the vN Clifford algebra $\mathcal{A}[L^\perp]$ when L is the linear span of any unit vector $l \in V$. Indeed, recall from Theorem 1.1.18 that l determines an isomorphism of involutive algebras from $C(L^\perp)$ to $C^+(V)$. An elementary modification of the argument surrounding Theorem 1.3.8 shows that this extends to an isometric isomorphism $U : \mathbb{H}(L^\perp) \to \mathbb{H}^+(V)$ and that the resulting map

$$B\big(\mathbb{H}(L^\perp)\big) \to B\big(\mathbb{H}^+(V)\big) : A \mapsto UAU^*$$

restricts to an isomorphism

$$\mathcal{A}[L^\perp] \to \mathcal{A}^+[V]$$

when the alternative description of $\mathcal{A}^+[V]$ in Theorem 1.3.10 is taken into account. Since we shall not have occasion to refer to this analogue of Theorem 1.2.12 later, we shall not pause to formalize it. However, we shall record two consequences for the structure of the even vN Clifford algebra.

Theorem 1.3.11 *The von Neumann algebras $\mathcal{A}^+[V]$ and $\mathcal{A}[V]$ are isomorphic.*

Proof Theorem 1.3.2 grants us the right to suppose that V is a Hilbert space. If L is any line in V then $\mathcal{A}^+[V]$ and $\mathcal{A}[L^\perp]$ are isomorphic, as we have just seen. Moreover, $\mathcal{A}[L^\perp]$ and $\mathcal{A}[V]$ are isomorphic: indeed, $L^\perp$ and V are equidimensional and if $g : L^\perp \to V$ is an isometric isomorphism then the resulting $\theta_g : C(L^\perp) \to C(V)$ extends to an isometric isomorphism $\mathbb{H}(L^\perp) \to \mathbb{H}(V)$ conjugation by which maps $\mathcal{A}[L^\perp]$ to $\mathcal{A}[V]$ after the familiar pattern.

$\square$

Theorem 1.3.12 $\mathcal{A}^+[V]$ *is a type II_1 factor.*

Proof This follows immediately from either Theorem 1.3.11 or the discussion preceding it, in view of Theorem 1.3.6 and Theorem 1.3.7.

$\square$

The factorial nature of $\mathcal{A}^+[V]$ enables us to show that the grading automorphism γ of $\mathcal{A}[V]$ is not inner, extending to the vN Clifford algebra earlier results with similar effect for the C^* Clifford algebra and the complex Clifford algebra. The proof of Theorem 1.1.20 adapts directly to handle the vN Clifford algebra; we record the result for later reference.

Theorem 1.3.13 *The grading automorphism γ of the vN Clifford algebra $\mathcal{A}[V]$ is not inner.* $\square$

More generally, deciding precisely which Bogoliubov automorphisms of the vN Clifford algebra are inner is a task that will be taken up and laid to rest in Section 4.3.

A natural problem at once presents itself when faced with a von Neumann algebra $\mathcal{A} \subset B(\mathbb{H})$: namely the determination of its commutant, this being the von Neumann algebra $\mathcal{A}'$ comprising all bounded linear operators on $\mathbb{H}$ that commute with each element of $\mathcal{A}$. The remainder of this section will be devoted to a solution of this problem for the vN Clifford algebra. It will be convenient to introduce some alternative notation: in place of $\mathcal{A}[V]$ we shall write $\mathcal{A}_\lambda[V]$ or simply $\mathcal{A}_\lambda$ to indicate that the vN Clifford algebra is generated by the left regular representation λ.

In view of the fact that multiplications on left and right commute, we are led to study the von Neumann algebra generated by right multiplications. Thus, consider the map

$$\rho : C(V) \rightarrow \operatorname{End} H_\tau$$

given by

$$a \in C(V), \ \zeta \in H_\tau \quad \Rightarrow \quad \rho(a)\zeta = \zeta a.$$

This map ρ is an antirepresentation: although linear, it does not preserve products but rather reverses them; we shall refer to it as the *right regular antirepresentation*. As with the left regular representation, ρ automatically extends to a star-antirepresentation

$$\rho : C[V] \rightarrow B(\mathbb{H}_\tau)$$

which we continue to call the right regular antirepresentation. We denote by $\mathcal{A}_\rho[V]$, or simply $\mathcal{A}_\rho$, the von Neumann algebra on $\mathbb{H}_\tau$ generated by either $\rho\big(C(V)\big)$ or $\rho\big(C[V]\big)$: thus, we put

$$\mathcal{A}_\rho := \rho\big(C(V)\big)'' \subset B(\mathbb{H}_\tau).$$

As we shall see, it transpires that the commutant of $\mathcal{A}_\lambda$ is none other than $\mathcal{A}_\rho$.

It is perhaps worth pointing out that the situation as regards commutants within $\operatorname{End} H_\tau$ is quite elementary.

Theorem 1.3.14 *Taking commutants within* $\operatorname{End} H_\tau$ *yields*

$$\lambda\big(C(V)\big)' = \rho\big(C(V)\big)$$
$$\rho\big(C(V)\big)' = \lambda\big(C(V)\big).$$

Proof We give arguments in support of the first equality, the second being entirely similar. Since multiplications on the left and right commute, the inclusion $\rho\big(C(V)\big) \subset \lambda\big(C(V)\big)'$ is plain. For the reverse inclusion, let $T \in \operatorname{End} H_\tau$ commute with $\lambda(a)$ whenever $a \in C(V)$ and put $b = T(\Omega) \in C(V)$. If $\zeta \in H_\tau$ then

$$\begin{aligned}
T(\zeta) = T(\zeta\Omega) &= T(\lambda(\zeta)\Omega) \\
&= \lambda(\zeta)T(\Omega) = \zeta b \\
&= \rho(b)\zeta
\end{aligned}$$

so that $T = \rho(b)$ lies in $\rho(C(V))$.

$\square$

Notice that this is of course valid for the left regular representation and right regular antirepresentation of any unital algebra on itself.

Concerning commutants within $B(\mathbb{H}_\tau)$ itself, we of course have the following result.

Theorem 1.3.15 $\mathcal{A}_\rho \subset \mathcal{A}'_\lambda$ and $\mathcal{A}_\lambda \subset \mathcal{A}'_\rho$.

Proof The fact that multiplications on left and right commute implies that $\rho(C(V)) \subset \lambda(C(V))'$ when commutants are taken in $B(\mathbb{H}_\tau)$. The taking of bicommutants now yields

$$\mathcal{A}_\rho = \rho(C(V))'' \subset \lambda(C(V))''' = \mathcal{A}'_\lambda.$$

The inclusion $\mathcal{A}_\lambda \subset \mathcal{A}'_\rho$ follows either by a similar argument or by taking commutants once more.

$\square$

Earlier we made an announcement to the effect that these inclusions are in fact equalities. We now introduce an operator that helps fulfill the promise of this announcement. Let us denote the main involution on $C(V) = H_\tau$ by Σ: thus, if $\zeta \in H_\tau$ then $\Sigma(\zeta) = \zeta^*$. Of course, Σ is both antilinear and of period 2; moreover, if $\xi, \eta \in H_\tau$ then

$$\begin{aligned}
\langle \Sigma\xi \mid \Sigma\eta \rangle &= \langle \xi^* \mid \eta^* \rangle \\
&= \tau(\eta\,\xi^*) \\
&= \tau(\xi^*\eta) \\
&= \langle \eta \mid \xi \rangle
\end{aligned}$$

since τ is central. In consequence, Σ extends by continuity to an antiunitary operator of period 2 on $\mathbb{H}_\tau$. We shall continue the notation Σ for this operator and refer to it as the *modular involution*, for reasons to be made clear later.

Theorem 1.3.16 *If $A \in \mathcal{A}_\lambda$ then $\Sigma(A\Omega) = A^*\Omega$.*

Proof Suppose first that $a \in C(V)$ and compute: since λ is involution-preserving, we have

$$\Sigma\big(\lambda(a)\Omega\big) = \Sigma(a) = a^*$$
$$= \lambda(a^*)\Omega$$
$$= \lambda(a)^*\Omega.$$

Now let $A \in \mathcal{A}_\lambda$. If $(a_j : j \in \mathcal{J})$ is a net in $C(V)$ such that $\lambda(a_j) \xrightarrow{w} A$ and if $\zeta \in H_\tau$ then

$$\langle \Sigma(A\Omega) \mid \zeta \rangle = \langle \Sigma\zeta \mid A\Omega \rangle$$
$$= \lim_j \langle \Sigma\zeta \mid \lambda(a_j)\Omega \rangle$$
$$= \lim_j \langle \Sigma\big(\lambda(a_j)\Omega\big) \mid \zeta \rangle$$
$$= \lim_j \langle \lambda(a_j)^*\Omega \mid \zeta \rangle$$
$$= \lim_j \langle \Omega \mid \lambda(a_j)\zeta \rangle$$
$$= \langle \Omega \mid A\zeta \rangle = \langle A^*\Omega \mid \zeta \rangle.$$

Thus, $\Sigma(A\Omega) = A^*\Omega$ as claimed.

$\square$

Of course, a precisely similar argument shows that if $B \in \mathcal{A}_\rho$ then $\Sigma(B\Omega) = B^*\Omega$. In fact, it turns out that $\Sigma(T\Omega) = T^*\Omega$ whenever T lies in the commutant $\mathcal{A}'_\lambda \supset \mathcal{A}_\rho$ of $\mathcal{A}_\lambda$. To see that this is so, let $A \in \mathcal{A}_\lambda$ and note from Theorem 1.3.16 that

$$\langle \Sigma(T\Omega) \mid A\Omega \rangle = \langle \Sigma(A\Omega) \mid T\Omega \rangle$$
$$= \langle A^*\Omega \mid T\Omega \rangle$$
$$= \langle \Omega \mid AT\Omega \rangle$$
$$= \langle \Omega \mid TA\Omega \rangle$$
$$= \langle T^*\Omega \mid A\Omega \rangle$$

since T and A commute. An application of the fact that Ω is cyclic for $\mathcal{A}_\lambda$ now yields the desired conclusion that $\Sigma(T\Omega) = T^*\Omega$.

Theorem 1.3.17 *If $T \in \mathcal{A}'_\lambda$ then $\Sigma(T\Omega) = T^*\Omega$.* $\square$

The modular involution Σ has two further properties that conspire to produce the inclusions opposite to those in Theorem 1.3.15; we take each of them in turn.

The first property is relatively straightforward. We begin with the observation that if $a \in C(V)$ then

$$\Sigma \circ \lambda(a) = \rho(a^*) \circ \Sigma$$

on $\mathbb{H}_\tau$ since if also $\zeta \in H_\tau$ then

$$\Sigma \circ \lambda(a)(\zeta) = \Sigma(a\zeta)$$

$$= (a\zeta)^*$$
$$= \zeta^* a^*$$
$$= \rho(a^*) \circ \Sigma(\zeta).$$

The period of Σ being two and ρ preserving involutions, it follows that if $a \in C(V)$ then

$$\Sigma \circ \lambda(a) \circ \Sigma = \rho(a)^*.$$

Taking closures in the weak operator topology, relative to which involution is continuous, yields the following.

Theorem 1.3.18 $\Sigma \mathcal{A}_\lambda \Sigma \subset \mathcal{A}_\rho$. $\qquad\qquad\square$

The second property makes use of Theorem 1.3.17 and more ingenuity: we claim that $\mathcal{A}'_\lambda \subset \Sigma \mathcal{A}_\lambda \Sigma$. To see that this is the case, let R, S, T all lie in $\mathcal{A}'_\lambda$. From Theorem 1.3.17 it follows that

$$\Sigma R \Sigma (ST\Omega) = \Sigma(RT^* S^* \Omega)$$
$$= STR^* \Omega$$
$$= S\Sigma(RT^* \Omega)$$
$$= S\Sigma R\Sigma(T\Omega)$$

whence

$$(\Sigma R \Sigma)S = S(\Sigma R \Sigma)$$

since Ω is cyclic for $\mathcal{A}'_\lambda \supset \mathcal{A}_\rho$. As a result,

$$\Sigma \mathcal{A}'_\lambda \Sigma \subset \mathcal{A}''_\lambda = \mathcal{A}_\lambda$$

and so $\mathcal{A}'_\lambda \subset \Sigma \mathcal{A}_\lambda \Sigma$ as claimed since Σ has period 2.

Theorem 1.3.19 $\mathcal{A}'_\lambda \subset \Sigma \mathcal{A}_\lambda \Sigma$. $\qquad\qquad\square$

At last, we assemble the pieces that together identify the commutant of the vN Clifford algebra. We have the chain of inclusions

$$\mathcal{A}_\rho \subset \mathcal{A}'_\lambda \subset \Sigma \mathcal{A}_\lambda \Sigma \subset \mathcal{A}_\rho$$

in which the first is given by Theorem 1.3.15, the second by Theorem 1.3.19 and the third by Theorem 1.3.18. Each of these inclusions is therefore an equality, so that $\mathcal{A}'_\lambda = \Sigma \mathcal{A}_\lambda \Sigma = \mathcal{A}_\rho$. A symmetric argument, or the taking of commutants, shows that $\mathcal{A}'_\rho = \Sigma \mathcal{A}_\rho \Sigma = \mathcal{A}_\lambda$ also.

Theorem 1.3.20 $\mathcal{A}'_\lambda = \Sigma \mathcal{A}_\lambda \Sigma = \mathcal{A}_\rho$. $\qquad\qquad\square$

Thus, the commutant of the vN Clifford algebra $\mathcal{A}[V] = \mathcal{A}_\lambda[V]$ is precisely $\mathcal{A}_\rho[V]$: the von Neumann algebra generated by the right regular antirepresentation.

We draw this section to a close by explaining why we chose to call Σ the modular involution. If $\mathcal{A}$ is a von Neumann algebra on a Hilbert space $\mathbb{H}$ in which Ω is a separating cyclic vector, then the Tomita-Takesaki modular theory begins with a consideration of the operator S_0 densely-defined in $\mathbb{H}$ by sending $A\Omega$ to $A^*\Omega$ whenever $A \in \mathcal{A}$. This operator is closeable and its closure S has polar decomposition $S = T\Delta^{\frac{1}{2}}$ in which T is an antiunitary operator of period 2 and Δ is a (possibly unbounded) positive operator; it is customary to refer to Δ as the *modular operator* and to T as the *modular involution* (or conjugation) associated to the pair $(\mathcal{A}, \Omega)$. In the case of the vN Clifford algebra, Theorem 1.3.16 makes it plain that S is precisely Σ and so already a period 2 antiunitary operator, whence $\Delta = I$ and Σ is indeed the modular involution.

Remarks

Complex Clifford algebras

It is possible to extract from our presentation an essentially combinatorial construction of the complex Clifford algebra over an m-dimensional real inner product space. Specifically, we let C_m be the algebra of all complex-valued functions defined on the power set of $\mathbf{m} = \{1, \ldots, m\}$, with pointwise linear operations and with product given by

$$(a.b)(T) = \Sigma\{\varepsilon(R, S)a(R)b(S) : R\Delta S = T\}$$

for $a, b \in C_m$ and $T \subset \mathbf{m}$; here, ε is as defined prior to Theorem 1.1.5. In these terms, the grading automorphism and main involution are given by

$$\gamma(a)(S) = (-1)^{|S|}a(S)$$
$$a^*(S) = (-1)^{\frac{1}{2}|S|(|S|-1)} \overline{a(S)}$$

for $a \in C_m$ and $S \subset \mathbf{m}$, while the canonical trace becomes evaluation at the empty set:

$$a \in C_m \quad \Rightarrow \quad \tau(a) = a(\emptyset).$$

Recall that if V is an m-dimensional real inner product space having $\{v_1, \ldots, v_m\}$ as orthonormal basis, then each $a \in C(V)$ assumes the form $\sum_{S \subset \mathbf{m}} \mu_S v_S$; the rule $S \subset \mathbf{m} \Rightarrow a(S) = \mu_S$ identifies $C(V)$ with C_m. Variants of this approach to Clifford algebras may be found in [5] [53] [89] for example.

C* Clifford algebras

The literature abounds in alternative formulations of C^* Clifford algebras.

First of all, let W be a complex Hilbert space provided with a conjugation operator Σ: thus, Σ is antiunitary of period 2. The associated *self-dual CAR algebra* $C[W, \Sigma]$ is the unital complex C^* algebra generated by $\{b(w) : w \in W\}$ where $b : W \to C[W, \Sigma]$ is complex-linear, subject to the self-duality relations

$$w \in W \quad \Rightarrow \quad b(\Sigma w) = b(w)^*$$

and the (partial) anticommutation relations

$$b(x)b(y)^* + b(y)^*b(x) = \langle x \mid y \rangle \mathbf{1}$$

for $x, y \in W$. Contact with the C^* Clifford algebra is made as follows: $\langle \cdot \mid \cdot \rangle$ restricts to a complete real inner product $(\cdot \mid \cdot)$ on the real subspace V of W fixed by Σ pointwise; in the opposite direction, W arises from V by complexification. Now, the rule

$$V \to C[W, \Sigma] : v \mapsto \tfrac{1}{\sqrt{2}} b(v)$$

defines a self-adjoint Clifford map inducing an isomorphism $C[V] \to C[W, \Sigma]$. This self-dual CAR algebra approach is adopted by Araki, among others: for example, see [1] [2] [3].

Incidentally, if we drop the conjugation operator Σ and replace the self-duality relations by the remaining canonical anticommutation relations

$$x, y \in W \quad \Rightarrow \quad b(x)b(y) + b(y)b(x) = 0$$

together with their adjoints, then we obtain simply the *CAR algebra* of the complex Hilbert space W. For more on this, see [2] [15] [65] [66].

Now suppose that the infinite-dimensional real Hilbert space V is separable. Let $\{x_n, y_n : n > 0\}$ be a complete orthonormal system for V and when $n > 0$ let V_n be the linear span of $\{x_1, y_1, \ldots, x_n, y_n\}$. The union of the increasing chain $C(V_1) \subset C(V_2) \subset \cdots$ is of course dense in $C[V]$; moreover, if $n > 0$ then the C^* algebra $C(V_n)$ is star-isomorphic to the full algebra of $2^n \times 2^n$ complex matrices, as noted under "Spin representations" in the Remarks at the end of Chapter Two. Thus: the C^* Clifford algebra $C[V]$ is a UHF (uniformly hyperfinite; also uniformly matricial or Glimm) C^* algebra of type $\{2^n : n > 0\}$. Explanations of this description may be found in many places: for example, see [50] (Chapter 10) or [61] (Chapter 6). In this connection, we remark that these C^* Clifford algebras and CAR algebras are often also constructed as countably-infinite tensor products of matrix algebras (involving the

Jordan-Wigner isomorphism) or from countably-infinite families of commuting 2×2 matrix algebras; see [42] [65].

vN Clifford algebras

Much more can be said about the vN Clifford algebra $\mathcal{A}[V]$ when V is an infinite-dimensional real Hilbert space. In Theorem 1.3.6 and Theorem 1.3.7 we demonstrated that $\mathcal{A}[V]$ is a factor of type II_1. An exposition of the classification of factors due to Murray and von Neumann is given in [50] starting at Chapter 6; see also [32] [84] [87] for example. The canonical trace τ on $\mathcal{A}[V]$ in Theorem 1.3.3 plays the rôle of a dimension function: if E is a projection in $\mathcal{A}[V]$ then $\tau(E)$ may be regarded as the dimension of E (or of its range). Here the range of τ is the entire unit interval $[0,1]$ so that the space of projections in $\mathcal{A}[V]$ forms a continuous complex projective geometry in the sense of von Neumann [59]. Detailed accounts of traces and dimension functions in the general context of von Neumann algebras may be found in Chapter 8 of [50]; see also [32] [84] [85]. If we suppose additionally that the infinite-dimensional real Hilbert space V is separable then the type II_1 factor $\mathcal{A}[V]$ is *hyperfinite*: the union of the increasing chain $C(V_1) \subset C(V_2) \subset \cdots$ of matrix algebras introduced under "C^* Clifford algebras" above is (weakly or strongly) dense in $\mathcal{A}[V]$. As a matter of fact, the hyperfinite II_1 factor is known to be unique up to isomorphism; it arises in a number of forms, each of which offers different insights on its structure. See [50] (Chapters 6, 8, 12) and [61] (Chapter 4) for example.

Conventions

Throughout our account, we have adopted the rule

$$v \in V \quad \Rightarrow \quad v^2 = (v \mid v)\mathbf{1}$$

as the fundamental Clifford property. It should be mentioned that the mathematical community is divided as to whether to adopt this or the alternative convention according to which $v^2 = -(v \mid v)\mathbf{1}$ when $v \in V$. Each convention has a substantial following among the authorities. We merely point out here that under the alternative convention, we would be led to define involution on the Clifford algebra by stipulating that it restrict to V as minus the identity, in order that $v^* v$ be positive when $v \in V$. Thus, the alternative convention would require elements of V to be skew-adjoint rather than self-adjoint in the Clifford algebra. Related comments regarding conventions will be found in the Remarks at the end of Chapter Four.

History and miscellany

Clifford algebras were of course introduced by William Kingdon Clifford (1845–1879) who wrote two papers on the subject: "Applications of Grassmann's extensive algebra" [26] and an unfinished manuscript "On the classification of geometric algebras" dating from 1876. Clifford himself considered an algebra generated by units $\iota_1, \ldots, \iota_n$ subject to $\iota_r^2 = -1$ and $\iota_s\iota_t + \iota_t\iota_s = 0$: the alternative convention! Historical remarks on Clifford algebras are to be found in [30] and [88]. Chevalley presents a rather detailed account of finite-dimensional Clifford algebras over arbitrary fields in [24]. Applications of Clifford algebras to physics stimulated their study: they enter fundamentally into the theory of electrons, witness Dirac [31] and Jordan & Wigner [49], for example. Applications within mathematics itself likewise stimulated growth: for example, see Brauer & Weyl [16] on spin representations and Atiyah, Bott & Shapiro [6] on K-theory.

The development of C^* Clifford algebras and vN Clifford algebras is intimately linked to that of quantum field theory and quantum statistical mechanics. These algebras appear in work of Segal [78]: he refers to a representative of the ring of Clifford distributions over a Hilbert space. The vN Clifford algebra also features in Blattner [13] underlying the construction of group representations by outer automorphisms. The work of Shale & Stinespring [80] [81] on the states and representations of C^* Clifford algebras served to herald a particularly fertile period in the development of the theory. This period continues to the present day; we shall sample some of its fruit later.

Finally, a couple of closing remarks. The result (Theorem 1.2.13) that the even C^* Clifford algebra is isomorphic to the C^* Clifford algebra itself was first established by Størmer [82] in the specific context of CAR algebras over separable complex Hilbert spaces and the original proof made use of matrix units. The rather different proof offered in the text is quite recent, appearing in [70]. The idea of using conditional expectations to establish the technical Theorem 1.2.20 is taken from Lemma 4.10 in Araki [3]; the original argument is modified and simplified somewhat for our purposes. Our identification of the commutant of the vN Clifford algebra falls within the domain of the standard Tomita-Takesaki modular theory: for details on this, refer to Chapter 9 of [50] for example.

2

FOCK REPRESENTATIONS

In this second chapter, we take a real Hilbert space V and study what are arguably the most important representations of its C^* Clifford algebra $C[V]$: namely, its Fock representations. For their definition, these Fock representations require that V be converted into a complex Hilbert space V_J by the introduction of a unitary structure J: that is, an orthogonal transformation of V whose square is minus the identity; this restricts the dimension of V to be other than odd. The Fock representation π_J of $C[V]$ determined by J will here be fashioned from creators and annihilators on Fock space, this being the complex Hilbert space completion of the complex exterior algebra $\bigwedge(V_J)$ relative to a canonical inner product. Alternatively, the Fock representation π_J may be usefully described (up to equivalence) as the unique representation π of $C[V]$ having a cyclic vector Ω with the J-vacuum property that if $v \in V$ then $\pi(v + \mathrm{i}Jv)\Omega = 0$. Incidentally if $v \in V$ then the creator $c(v) = c_J(v)$ and annihilator $a(v) = a_J(v)$ may be recovered from $\pi = \pi_J$ by the formulae $c(v) = \frac{1}{2}\pi(v - \mathrm{i}Jv)$ and $a(v) = \frac{1}{2}\pi(v + \mathrm{i}Jv)$. These creators and annihilators satisfy the famous canonical anticommutation relations: if $x, y \in V$ then

$$c(x)a(y) + a(y)c(x) = \langle x \mid y \rangle I$$
$$c(x)c(y) + c(y)c(x) = 0$$
$$a(x)a(y) + a(y)a(x) = 0$$

where $\langle \cdot \mid \cdot \rangle$ is the inner product on V_J as a complex Hilbert space.

In §1 we discuss carefully the notion of a unitary structure, considering it from three different viewpoints each of which has its merits. In

§2 we present a construction of the Fock space $\mathbb{H}_J(V)$ arising from a choice J of unitary structure; here, we develop norm estimates by which we construct Gaussians or quadratic exponentials. In §3 we introduce and investigate creators and annihilators on Fock space; in particular, we specify their kernels. The Fock representation π_J itself is taken up in §4: we see that it is irreducible and is determined up to equivalence either by the corresponding Fock vacuum or by the corresponding Fock state. Finally, in §5 we attend to matters of parity: upon restriction to the even C^* Clifford algebra, π_J decomposes as the direct sum of two inequivalent irreducible representations. The Remarks section with which this chapter closes contains additional information: among other things, we discuss Fock representations as spin representations and place them in the broader context of quasifree representations.

2.1 Unitary structures

Let V be a real Hilbert space: thus, suppose V now to be complete. In order to construct a Fock representation of the C^* Clifford algebra $C[V]$ it is first necessary to convert V into a complex Hilbert space. Our task in this section is to investigate the details of such a conversion. In fact, we shall consider three equivalent procedures by which such a conversion may be effected. The first and most frequently used of these procedures is purely internal: it is to introduce on V a unitary structure, meaning a complex structure adapted to the inner product. The alternative procedures are partly external, requiring us to extend from real to complex scalars and consider the complex Hilbert space $V^{\mathbb{C}}$: the one procedure is to decompose $V^{\mathbb{C}}$ as the orthogonal sum of two conjugate closed subspaces; the other is essentially to focus not on such a decomposition but rather on the corresponding projections. Without further delay, let us get on with our task and discuss each conversion procedure in detail.

By definition, a *unitary structure* on V is an orthogonal transformation $J \in O(V)$ with the property that $J^2 = -I$. The assumption that $J^2 = -I$ allows us to convert V into a complex vector space by stipulating that $iv = J(v)$ whenever $v \in V$. In the interests of clarity, we may write V_J for the vector space V made complex via J in this fashion. Given $x, y \in V$ we shall put

$$\langle x \mid y \rangle_J = (x \mid y) + \mathrm{i}(x \mid Jy).$$

This prescription defines a (positive definite) Hermitian inner product $\langle \cdot \mid \cdot \rangle_J$ on the complex vector space V_J. When only one unitary structure is under consideration and confusion is unlikely to result, we may write $\langle \cdot \mid \cdot \rangle$ rather than $\langle \cdot \mid \cdot \rangle_J$ for convenience. Note that $\langle \cdot \mid \cdot \rangle_J$ induces the same norm on V as does the original real inner product $(\cdot \mid \cdot)$. As a result, $\langle \cdot \mid \cdot \rangle_J$ actually makes V_J into a complex Hilbert space.

Theorem 2.1.1 *The real Hilbert space V admits unitary structures if and only if its dimension is other than odd.*

Proof Plainly, V cannot carry a unitary structure if its real dimension is odd. In the complementary case, let $\{x_j, y_j : j \in \mathcal{J}\}$ be a complete orthonormal system for V indexed in pairs. Define $J : V \to V$ by requiring that $Jx_j = y_j$ and $Jy_j = -x_j$ whenever $j \in \mathcal{J}$, extending by linearity and by continuity if V is infinite-dimensional. The resulting J is evidently a unitary structure on V. □

In the light of this result, we shall implicitly suppose that the dimension of V is either even or infinite until explicitly stated otherwise.

By the (Hilbert space) *dimension* of a (real or complex) Hilbert space we mean the cardinality of a complete orthonormal system. Having said this, let J be a unitary structure on V. As is readily apparent, if $\mathcal{C}$ is a complete orthonormal system for the complex Hilbert space V_J then $\mathcal{C} \cup J\mathcal{C}$ is a complete orthonormal system for the real Hilbert space V. Accordingly, we have the following result.

Theorem 2.1.2 *Let J be a unitary structure on V. If V is infinite-dimensional then the dimension of V_J equals that of V; if V is even-dimensional then the dimension of V_J equals half that of V.* □

Let us agree to write $\mathbb{U}(V)$ for the set of all unitary structures on V. The orthogonal group $O(V)$ acts on $\mathbb{U}(V)$ by conjugation: if $g \in O(V)$ and $J \in \mathbb{U}(V)$ then gJg^{-1} is orthogonal and has square equal to minus the identity. In fact, this action is transitive: indeed, if J and K are unitary structures on V then Theorem 2.1.2 asserts that the complex Hilbert spaces V_J and V_K are equidimensional and hence isometrically isomorphic; if $g : V_J \to V_K$ is an isometric isomorphism, then its complex-linearity yields $K = gJg^{-1}$ and its isometric nature ensures its orthogonality. Moreover, the stabilizer of a specific unitary structure $J \in \mathbb{U}(V)$ under this action of $O(V)$ is precisely the unitary group $U(V_J)$ of the complex Hilbert space V_J: indeed, $g \in O(V)$ preserves the imaginary part of $\langle \cdot \mid \cdot \rangle_J$ if and only if it commutes with J.

Theorem 2.1.3 *The orthogonal group $O(V)$ acts transitively on the* *set $\mathbb{U}(V)$ of unitary structures on V by conjugation, the stabilizer of $J \in \mathbb{U}(V)$ being the unitary group $U(V_J)$.* ☐

Thus, $\mathbb{U}(V)$ is naturally a homogeneous space for the orthogonal group $O(V)$ with unitary groups as stabilizers. This is rather familiar in finite dimensions: the set of unitary structures on $\mathbb{R}^{2n}$ with the standard inner product is customarily identified with the homogeneous space $O(2n)/U(n)$ in the usual notation.

In order to formulate alternative perspectives on unitary structures, we complexify V. Thus, we let $V^{\mathbb{C}} = \mathbb{C} \otimes V$ be the complex vector space obtained from V by extending scalars from $\mathbb{R}$ to $\mathbb{C}$; as is customary, we identify $v \in V$ with $1 \otimes v \in \mathbb{C} \otimes V$ so that elements of $V^{\mathbb{C}}$ assume the form $x + iy$ for $x, y \in V$. The real inner product $(\cdot \mid \cdot)$ on V extends by sesquilinearity to define a (positive definite) Hermitian inner product $\langle \cdot \mid \cdot \rangle$ on $V^{\mathbb{C}}$ according to the rule

$$\langle \mu \otimes x \mid \nu \otimes y \rangle = \mu \bar{\nu} (x \mid y)$$

for $x, y \in V$ and $\mu, \nu \in \mathbb{C}$. Note that there is no conflict of notation here: $V^{\mathbb{C}}$ is a subspace of $C(V) = H_\tau \subset \mathbb{H}_\tau$ and the inner product defined above is exactly the restriction to $V^{\mathbb{C}}$ of that defined by the trace τ on $C(V)$ as in Theorem 1.1.9 and Theorem 1.1.17. To see this, invoke sesquilinearity after observing that the Clifford relations yield

$$x, y \in V \quad \Rightarrow \quad \tau(yx) = (x \mid y)$$

since τ is a central state. In particular, note that

$$x, y \in V \quad \Rightarrow \quad \|x + iy\|^2 = \|x\|^2 + \|y\|^2$$

whence it is clear that $\langle \cdot \mid \cdot \rangle$ makes $V^{\mathbb{C}}$ into a complex Hilbert space. Conjugation of $V^{\mathbb{C}}$ over V will be denoted by either Σ or an upper bar, so that

$$\Sigma(x + iy) = x - iy = \overline{x + iy}$$

whenever $x, y \in V$. Again there is no conflict of notation: Σ is exactly the restriction to $V^{\mathbb{C}}$ of the modular involution on $\mathbb{H}_\tau$ introduced prior to Theorem 1.3.16. The formula

$$x, y \in V^{\mathbb{C}} \quad \Rightarrow \quad (x \mid y) = \langle x \mid \Sigma y \rangle$$

determines the unique extension of the original real inner product on V to a (nonsingular symmetric) complex-bilinear form on $V^{\mathbb{C}}$ also denoted $(\cdot \mid \cdot)$.

Now, let $J \in \mathbb{U}(V)$ be a unitary structure on V and note that it extends uniquely to a complex-linear automorphism of $V^{\mathbb{C}}$ denoted by J again. Of course, the equation $J^2 = -I$ continues to hold for this extension. Since J on $V^{\mathbb{C}}$ arises by complex-linear extension of an operator on V it commutes with the involution Σ. The operator J also lies in the unitary group $U(V^{\mathbb{C}})$ of $V^{\mathbb{C}}$ relative to the inner product $\langle \cdot \mid \cdot \rangle$: indeed, if $x, y \in V$ and $\mu, \nu \in \mathbb{C}$ then

$$\begin{aligned}
\langle J(\mu \otimes x) \mid J(\nu \otimes y) \rangle &= \langle \mu \otimes Jx \mid \nu \otimes Jy \rangle \\
&= \mu\bar{\nu}(Jx \mid Jy) = \mu\bar{\nu}(x \mid y) \\
&= \langle \mu \otimes x \mid \nu \otimes y \rangle
\end{aligned}$$

since J is orthogonal. From $J^2 = -I$ and $J \in U(V^{\mathbb{C}})$ it follows that J is skew-adjoint relative to $\langle \cdot \mid \cdot \rangle$. From $J\Sigma = \Sigma J$ and $J \in U(V^{\mathbb{C}})$ it follows that J lies in the (complex) orthogonal group $O(V^{\mathbb{C}})$ of $V^{\mathbb{C}}$ relative to $(\cdot \mid \cdot)$. Armed with these various properties of the operator J, we may now formulate the promised alternative means of introducing unitary structures.

For the first alternative approach to unitary structures, observe that from $J^2 = -I$ it follows that $V^{\mathbb{C}}$ splits as the direct sum of eigenspaces

$$V^{\mathbb{C}} = F_J^+ \oplus F_J^-$$

where J acts as $\pm i$ on $F_J^{\pm}$: indeed, if $w \in V^{\mathbb{C}}$ then

$$w = \tfrac{1}{2}(w - iJw) + \tfrac{1}{2}(w + iJw)$$

where $\tfrac{1}{2}(w \mp iJw) \in F_J^{\pm}$. Of course, the eigenspaces F_J^+ and F_J^- are closed because J is continuous. Also, the spaces F_J^+ and F_J^- are *conjugate* in the sense that Σ interchanges them: indeed, if $w \in F_J^{\pm}$ then

$$J(\Sigma w) = \Sigma(Jw) = \Sigma(\pm iw) = \mp i\Sigma w$$

since J and Σ commute. From the fact that J lies in $U(V^{\mathbb{C}})$ it follows that F_J^+ and F_J^- are *orthogonal* relative to $\langle \cdot \mid \cdot \rangle$: if $w^{\pm} \in F_J^{\pm}$ then

$$\langle w^+ \mid w^- \rangle = \langle Jw^+ \mid Jw^- \rangle = \langle iw^+ \mid -iw^- \rangle = -\langle w^+ \mid w^- \rangle$$

whence $\langle w^+ \mid w^- \rangle = 0$. Lastly, F_J^+ and F_J^- are *isotropic* for $(\cdot \mid \cdot)$ in the sense that $(\cdot \mid \cdot)$ vanishes on each: this is readily seen directly but also follows from above, via the relationship between $(\cdot \mid \cdot)$ and $\langle \cdot \mid \cdot \rangle$ mediated by Σ. Let us summarize these findings in the following form.

Theorem 2.1.4 *Let $J \in \mathbb{U}(V)$. If F_J^+ and F_J^- are the eigenspaces of J in $V^{\mathbb{C}}$ with eigenvalues $+i$ and $-i$ respectively, then $F_J^- = \Sigma F_J^+$ and*

$$V^{\mathbb{C}} = F_J^+ \oplus F_J^-$$

is an orthogonal decomposition into isotropic closed subspaces. $\square$

Conversely, let $V^{\mathbb{C}} = F^+ \oplus F^-$ be a decomposition into closed subspaces. We contend that of the three conditions

(a) F^+ and F^- are conjugate

(b) F^+ and F^- are orthogonal

(c) F^+ and F^- are isotropic

any two imply the third. Taking the implication (b)$\wedge$(c)$\Rightarrow$(a) first, we must show that if w lies in F^+ then $\overline{w} = \Sigma w$ lies in F^-. Write $\overline{w} = x+y$ with $x \in F^+$ and $y \in F^-$. From

$$\|x\|^2 = \langle x \mid x \rangle = \langle x \mid \overline{w} \rangle - \langle x \mid y \rangle$$
$$= (x \mid w) - \langle x \mid y \rangle = 0$$

it follows that $x = 0$ and so $\overline{w} = y \in F^-$ as required. The implications (c)$\wedge$(a)$\Rightarrow$(b) and (a)$\wedge$(b)$\Rightarrow$(c) are even more straightforward than this.

Now, let $V^{\mathbb{C}} = F^+ \oplus F^-$ be a decomposition with the above properties and define a complex-linear automorphism J of $V^{\mathbb{C}}$ by $J \mid F^{\pm} = \pm \mathrm{i}$ so that $J^2 = -I$. Since F^+ and F^- are conjugate, J commutes with Σ and so restricts to an automorphism of V which we denote by J also. Since F^+ and F^- are orthogonal, the complex J lies in $U(V^{\mathbb{C}})$ and so the real J lies in $O(V)$. Thus, J is actually a unitary structure on V.

Theorem 2.1.5   *If $V^{\mathbb{C}} = F^+ \oplus F^-$ is an orthogonal decomposition into conjugate closed subspaces, then the complex-linear automorphism of $V^{\mathbb{C}}$ acting on $F^{\pm}$ as $\pm \mathrm{i}$ restricts to a unitary structure $J \in \mathbb{U}(V)$.* $\square$

We hardly need mention that the correspondences between unitary structures and decompositions expressed in Theorems 2.1.4 and 2.1.5 are mutually inverse: in particular, if the decomposition $V^{\mathbb{C}} = F^+ \oplus F^-$ gives rise to the unitary structure J then $F_J^{\pm} = F^{\pm}$. Incidentally, since $F_J^- = \Sigma F_J^+$ we shall usually simplify notation by writing $F_J^+ = F_J$ so that $F_J^- = \overline{F_J}$ whenever J is a unitary structure on V.

For the second alternative approach to unitary structures, again let $J \in \mathbb{U}(V)$ be extended to $V^{\mathbb{C}}$ by complex-linearity and put

$$P_J = \tfrac{1}{2}(I - \mathrm{i}J) \in B(V^{\mathbb{C}}).$$

From $J^2 = -I$ it follows that P_J is *idempotent*: $P_J^2 = P_J$. From $J^* = -J$ it follows that P_J is *self-adjoint*: $P_J^* = P_J$. Thus, P_J is a *projection operator* on the complex Hilbert space $V^{\mathbb{C}}$. It has the further property that $P_J + \Sigma P_J \Sigma = I$ since J and Σ commute.

Theorem 2.1.6   *If $J \in \mathbb{U}(V)$ is a unitary structure on V then the formula*

$$P_J = \tfrac{1}{2}(I - \mathrm{i}J)$$

defines a projection on $V^{\mathbb{C}}$ such that

$$P_J + \Sigma P_J \Sigma = I.$$ $\square$

Conversely, let P be a projection on $V^{\mathbb{C}}$ such that $P + \Sigma P \Sigma = I$ and put

$$J = \mathrm{i}(2P - I).$$

A routine computation shows that $J^2 = -I$ follows from P being idempotent. Another such computation yields $J \in U(V^{\mathbb{C}})$ from P being self-adjoint. Lastly, the relation $P + \Sigma P \Sigma = I$ is readily seen to imply that Σ and J commute. The restriction of J to V is consequently a unitary structure, conveniently denoted by J again.

Theorem 2.1.7 *If P is a projection on $V^{\mathbb{C}}$ such that $P + \Sigma P \Sigma = I$ then the restriction of $\mathrm{i}(2P - I)$ to V is a unitary structure $J \in \mathbb{U}(V)$.*

$\square$

Again, we need hardly mention that the correspondences between unitary structures and projections expressed in Theorem 2.1.6 and Theorem 2.1.7 are mutually inverse: in particular, if the projection P gives rise to the unitary structure J then $P_J = P$.

By transitivity, the decomposition approach and the projection approach to unitary structures are naturally equivalent: the equation $F = \operatorname{ran} P$ sets up a bijection between the set of closed subspaces $F \subset V^{\mathbb{C}}$ for which $V^{\mathbb{C}} = F \oplus \overline{F}$ is an orthogonal decomposition and the set of projections P on $V^{\mathbb{C}}$ such that $P + \Sigma P \Sigma = I$; under this bijection, P_J corresponds to F_J whenever $J \in \mathbb{U}(V)$.

In closing this section, it is convenient to record for later use one further property of unitary structures.

Theorem 2.1.8 *If $J \in \mathbb{U}(V)$ is a unitary structure on V then the map*

$$V_J \to F_J : v \mapsto \tfrac{1}{\sqrt{2}}(v - \mathrm{i}Jv)$$

is an isometric isomorphism.

Proof Here, we regard V_J as endowed with the inner product $\langle \cdot \mid \cdot \rangle_J$ and F_J as endowed with the restriction of $\langle \cdot \mid \cdot \rangle$ from $V^{\mathbb{C}}$. The indicated map being plainly a complex-linear isomorphism, all we need note is that if $x, y \in V$ then

$$\left\langle \tfrac{1}{\sqrt{2}}(x - \mathrm{i}Jx) \mid \tfrac{1}{\sqrt{2}}(y - \mathrm{i}Jy) \right\rangle = \langle x \mid y \rangle_J$$

and this follows by direct calculation from the definitions. $\square$

2.2 Fock spaces

Throughout the whole of this section, J will be a fixed choice of unitary structure on the real Hilbert space V whose dimension is necessarily other than odd. Since no confusion is likely to result, we shall denote the corresponding (positive definite) Hermitian inner product on V_J simply by $\langle \cdot \mid \cdot \rangle$ for convenience: thus

$$x, y \in V \quad \Rightarrow \quad \langle x \mid y \rangle = (x \mid y) + \mathrm{i}(x \mid Jy).$$

Our primary objective in this section is to describe the complex Hilbert space that will ultimately carry the Fock representation of the C^* Clifford algebra $C[V]$ determined by the unitary structure J.

We begin with the familiar fact that if n is any positive integer then the n-fold (complex) alternating power $\bigwedge^n(V_J)$ carries a canonical Hermitian inner product given on decomposables by the formula

$$\langle x_1 \wedge \ldots \wedge x_n \mid y_1 \wedge \ldots \wedge y_n \rangle = \mathrm{Det}[\,\langle x_i \mid y_j \rangle\,]$$

for $x_1, y_1, \ldots, x_n, y_n \in V$. The resulting inner product is incomplete if $n > 1$ and V is infinite-dimensional; we denote the Hilbert space completion by $\bigwedge^n[V_J]$ in all cases. Additionally, we write $\bigwedge^0[V_J] = \bigwedge^0(V_J)$ for $\mathbb{C}$ with its standard Hermitian inner product and $\Omega_J = 1$ for its standard unit vector.

We continue by forming the orthogonal sum of the inner product spaces $\bigwedge^n(V_J)$ as n runs over the non-negative integers. Thus

$$H_J = H_J(V) = \bigoplus_{n \geq 0} \bigwedge^n(V_J)$$

has elements $\bigoplus_{n \geq 0} \zeta_n$ where $\zeta_n \in \bigwedge^n(V_J)$ is zero for all but finitely many values of $n \geq 0$; the inner product between its elements $\bigoplus_{n \geq 0} \xi_n$ and $\bigoplus_{n \geq 0} \eta_n$ is given by

$$\left\langle \bigoplus_{n \geq 0} \xi_n \mid \bigoplus_{n \geq 0} \eta_n \right\rangle = \sum_{n \geq 0} \langle \xi_n \mid \eta_n \rangle.$$

The inner product space $H_J = H_J(V)$ is incomplete if V is infinite-dimensional and is finite-dimensional otherwise. Its Hilbert space completion is the *Fock space* $\mathbb{H}_J = \mathbb{H}_J(V)$ over V determined by the unitary structure J. Of course, $\mathbb{H}_J(V)$ contains (isometric copies of) the Hilbert spaces $\bigwedge^n[V_J]$ and indeed may be regarded as their Hilbert space sum over $n \geq 0$.

We remark that if $\{v_r : r \in \mathcal{R}\}$ is a complete orthonormal system for the complex Hilbert space V_J with totally ordered index set $\mathcal{R}$ then the collection of all vectors $v_{r_1} \wedge \ldots \wedge v_{r_n}$ with $r_1 < \ldots < r_n$ in $\mathcal{R}$ constitutes a complete orthonormal system for $\bigwedge^n[V_J]$ whenever $n \geq 0$

and the union of these collections is a complete orthonormal system for the Fock space $\mathbb{H}_J(V)$ itself; here, if $n = 0$ then the indicated alternating product is understood to mean Ω_J. We shall refer to $R = \{r_1, \ldots, r_n\}$ with $r_1 < \ldots < r_n$ in $\mathcal{R}$ as a (strictly increasing) *multiindex* from $\mathcal{R}$ and shall then write $R \uparrow \mathcal{R}$. In case $R = \{r_1 < \ldots < r_n\} \uparrow \mathcal{R}$ we shall put

$$v_R = v_{r_1} \wedge \ldots \wedge v_{r_n}$$

with the understanding that $v_\emptyset = \Omega_J$. Although this duplicates the abbreviated notation for specific elements of Clifford algebras, any potential ambiguity will be clearly resolved by context. Thus, the set $\{v_R : R \uparrow \mathcal{R}\}$ is a complete orthonormal system for Fock space $\mathbb{H}_J(V)$. In particular, each $\zeta \in \mathbb{H}_J(V)$ has a (generalized) Fourier expansion

$$\zeta = \sum_{R \uparrow \mathcal{R}} \langle \zeta \mid v_R \rangle v_R.$$

Recall that the wedge product $\wedge$ converts the complex vector space $\bigoplus_{n \geq 0} \bigwedge^n(V_J)$ into an associative complex algebra: the *exterior algebra* (or Grassmann algebra) $\bigwedge(V_J)$ over V_J. Thus

$$H_J(V) = \bigoplus_{n \geq 0} \bigwedge^n(V_J) = \bigwedge(V_J).$$

Now, the wedge product does not extend continuously to a multiplication on the Fock space $\mathbb{H}_J(V)$ when V is infinite-dimensional. However, it so happens that if non-negative integers p and q are fixed then

$$\wedge : \bigwedge^p(V_J) \times \bigwedge^q(V_J) \to \bigwedge^{p+q}(V_J)$$

does extend continuously to define a wedge product

$$\wedge : \bigwedge^p[V_J] \times \bigwedge^q[V_J] \to \bigwedge^{p+q}[V_J]$$

with the property that if $\xi \in \bigwedge^p[V_J]$ and $\eta \in \bigwedge^q[V_J]$ then

$$\|\xi \wedge \eta\|^2 \leq \frac{(p+q)!}{p!q!} \|\xi\|^2 \|\eta\|^2.$$

We proceed to establish this last fact, although our main interest actually lies in a refinement of a special case.

First of all, fix a complete orthonormal system $\{v_r : r \in \mathcal{R}\}$ for V_J as above. Let $\xi \in \bigwedge^p(V_J)$ and $\eta \in \bigwedge^q(V_J)$ so that the Fourier expansions

$$\xi = \sum_{R \uparrow \mathcal{R}} \xi_R v_R \quad , \quad \eta = \sum_{R \uparrow \mathcal{R}} \eta_R v_R$$

are finite sums. Now,

$$\xi \wedge \eta = \sum_{S, T \uparrow \mathcal{R}} \xi_S \, \eta_T \, v_S \wedge v_T$$

where summation need only be performed over the multiindices S and T that are disjoint, since the alternating nature of the wedge product forces $v_S \wedge v_T = 0$ when $S \cap T \neq \emptyset$. The alternating nature of the wedge product also implies that if $R = S \cup T$ with $S \cap T = \emptyset$ then

$$v_R = \pm v_S \wedge v_T$$

where $\pm$ is the sign of the permutation that effects the resorting of $S \cup T$ into increasing order. Thus

$$\xi \wedge \eta = \sum_{R \uparrow \mathcal{R}} \zeta_R v_R$$

where if $R \uparrow \mathcal{R}$ then

$$\zeta_R = \sum_{R = S \cup T} (\pm \xi_S \eta_T)$$

where summation implicitly involves disjoint multiindices S and T of cardinalities p and q respectively. Since a given multiindex of cardinality $p + q$ has $(p+q)!/p!q!$ decompositions as the union of multiindices having cardinalities p and q, it follows from the (classical) Cauchy-Schwarz inequality that if $R \uparrow \mathcal{R}$ then

$$|\zeta_R|^2 \leq \frac{(p+q)!}{p!q!} \sum_{R = S \cup T} |\xi_S|^2 |\eta_T|^2.$$

The Parseval equality now implies that

$$\|\xi \wedge \eta\|^2 = \sum_{R \uparrow \mathcal{R}} |\zeta_R|^2$$

$$\leq \frac{(p+q)!}{p!q!} \sum_{R \uparrow \mathcal{R}} \sum_{R = S \cup T} |\xi_S|^2 |\eta_T|^2$$

and taking into account intersecting multiindices yields

$$\sum_{R \uparrow \mathcal{R}} \sum_{R = S \cup T} |\xi_S|^2 |\eta_T|^2 \leq \left(\sum_{S \uparrow \mathcal{R}} |\xi_S|^2 \right) \left(\sum_{T \uparrow \mathcal{R}} |\eta_T|^2 \right)$$

$$= \|\xi\|^2 \|\eta\|^2.$$

Putting together the pieces, we have shown that if $\xi \in \bigwedge^p(V_J)$ and $\eta \in \bigwedge^q(V_J)$ then

$$\|\xi \wedge \eta\|^2 \leq \frac{(p+q)!}{p!q!} \|\xi\|^2 \|\eta\|^2.$$

Having reached this point, we need take only a couple of short steps. The inequality just established ensures that if $\xi \in \bigwedge^p[V_J]$ and $\eta \in \bigwedge^q[V_J]$ then we may unambiguously define

$$\xi \wedge \eta := \lim_n (\xi_n \wedge \eta_n) \in \bigwedge^{p+q}[V_J]$$

for any sequences $(\xi_n : n > 0)$ and $(\eta_n : n > 0)$ in $\bigwedge^p(V_J)$ and $\bigwedge^q(V_J)$ such that $\xi_n \to \xi$ and $\eta_n \to \eta$ respectively; continuity then guarantees that the inequality

$$\|\xi \wedge \eta\|^2 \le \frac{(p+q)!}{p!q!} \|\xi\|^2 \|\eta\|^2$$

continues to hold in this context. In short, the following result is both sensible and valid.

Theorem 2.2.1 *If $\xi \in \bigwedge^p[V_J]$ and $\eta \in \bigwedge^q[V_J]$ then*

$$\|\xi \wedge \eta\|^2 \le \frac{(p+q)!}{p!q!} \|\xi\|^2 \|\eta\|^2$$

$\square$

Of course, induction now implies that if $\zeta_j \in \bigwedge^{p_j}[V_J]$ for $j = 1, \dots, n$ then

$$\|\zeta_1 \wedge \dots \wedge \zeta_n\|^2 \le \frac{(p_1 + \dots + p_n)!}{p_1! \dots p_n!} \|\zeta_1\|^2 \dots \|\zeta_n\|^2.$$

In particular, if $\zeta \in \bigwedge^2[V_J]$ then the n-fold power $\zeta^n = \zeta \wedge \dots \wedge \zeta \in \bigwedge^{2n}[V_J]$ satisfies

$$\|\zeta^n\|^2 \le \frac{(2n)!}{2^n} \|\zeta\|^{2n}.$$

This inequality is the special case to which we alluded earlier. The refinement also alluded to is this: that if $\zeta \in \bigwedge^2[V_J]$ then in fact

$$\|\zeta^n\|^2 \le n! \|\zeta\|^{2n}.$$

We require this stronger inequality in order to guarantee convergence in Fock space $\mathbb{H}_J(V)$ of the exponential series

$$\exp \zeta = \sum_{n \ge 0} \frac{1}{n!} \zeta^n$$

whenever $\zeta \in \bigwedge^2[V_J]$. These *quadratic exponentials* (or *Gaussians*) will ultimately play an important rôle in our subsequent investigations into the unitary implementation of Bogoliubov automorphisms in the Fock representation.

Our approach to the stronger inequality is via an alternative description of $\bigwedge^2[V_J]$. Thus, let $\zeta \in \bigwedge^2[V_J]$ and define an alternating bilinear form

$$T_\zeta : V \times V \to \mathbb{C}$$

by

$$x, y \in V \quad \Rightarrow \quad T_\zeta(x, y) = \langle x \wedge y \mid \zeta \rangle.$$

Note that if $x, y \in V$ then

$$|T_\zeta(x, y)| \le \|x \wedge y\| \, \|\zeta\|$$
$$\le \|x\| \, \|y\| \, \|\zeta\|$$

in consequence of which inequality there exists a bounded real-linear operator $Z = Z_\zeta$ on V such that

$$x, y \in V \quad \Rightarrow \quad \langle Zx \mid y \rangle_J = \langle \zeta \mid x \wedge y \rangle.$$

The alternating nature of T_ζ implies that Z satisfies the condition

$$x, y \in V \quad \Rightarrow \quad \langle Zx \mid y \rangle + \langle Zy \mid x \rangle = 0$$

which in turn implies and is implied by J-antilinearity of Z together with the property

$$x, y \in V \quad \Rightarrow \quad (Zx \mid y) + (Zy \mid x) = 0.$$

It turns out that if $\zeta \in \bigwedge^2[V_J]$ then $Z = Z_\zeta$ is more than just bounded: it is actually of Hilbert-Schmidt class. To see that this is so, note that if $\{v_r : r \in \mathcal{R}\}$ is a complete orthonormal system for the complex Hilbert space V_J then $\{v_r, Jv_r : r \in \mathcal{R}\}$ is a complete orthonormal system for the real Hilbert space V; thus the (real) Hilbert-Schmidt norm of Z is given by

$$\begin{aligned}
\|Z\|_{\mathrm{HS}}^2 &= \sum_{r \in \mathcal{R}} \left\{ \|Zv_r\|^2 + \|ZJv_r\|^2 \right\} \\
&= 2 \sum_{s,t} |\langle Zv_s \mid v_t \rangle|^2 \\
&= 2 \sum_{s,t} |\langle \zeta \mid v_s \wedge v_t \rangle|^2 \\
&= 4 \sum_{s<t} |\langle \zeta \mid v_s \wedge v_t \rangle|^2 \\
&= 4\|\zeta\|^2
\end{aligned}$$

since $\bigwedge^2[V_J]$ has $\{v_s \wedge v_t : s < t \in \mathcal{R}\}$ as a complete orthonormal system.

Taking stock of our progress, let us denote by $\mathbb{S}(V_J)$ the vector space of all real Hilbert-Schmidt operators Z on V that are *antiskew* in satisfying the condition

$$x, y \in V \quad \Rightarrow \quad \langle Zx \mid y \rangle + \langle Zy \mid x \rangle = 0.$$

As noted above, antiskew operators necessarily anticommute with J. Nevertheless, $\mathbb{S}(V_J)$ is a complex vector space: if $Z \in \mathbb{S}(V_J)$ then $(a + bi)Z := aZ + bJZ$ whenever $a, b \in \mathbb{R}$. Still more is true: $\mathbb{S}(V_J)$ is in fact a complex Banach space relative to the real Hilbert-Schmidt norm. Thus far, we have constructed a complex linear map

$$\bigwedge^2[V_J] \to \mathbb{S}(V_J) : \zeta \mapsto Z_\zeta$$

with the property that $\|Z_\zeta\|_{\mathrm{HS}} = 2\|\zeta\|$ whenever $\zeta \in \bigwedge^2[V_J]$. This map is actually surjective: indeed, if $Z \in \mathbb{S}(V_J)$ is given then

$$\zeta := \sum_{s<t} \langle Zv_s \mid v_t \rangle \, v_s \wedge v_t$$

converges to define an element $\zeta \in \bigwedge^2[V_J]$ such that $Z = Z_\zeta$. We have arrived at the following alternative description of $\bigwedge^2[V_J]$.

Theorem 2.2.2 *A canonical isometric isomorphism from $\bigwedge^2[V_J]$ to $\mathbb{S}(V_J)$ assigns to $\zeta \in \bigwedge^2[V_J]$ the Hilbert-Schmidt antiskew operator $\frac{1}{2}Z_\zeta$ defined by*

$$x, y \in V \quad \Rightarrow \quad \langle Z_\zeta x \mid y \rangle = \langle \zeta \mid x \wedge y \rangle.$$

$\square$

Now, let $Z \in \mathbb{S}(V_J)$. The bounded complex-linear operator Z^2 on V_J is self-adjoint: indeed, if $x, y \in V$ then

$$\langle Z^2 x \mid y \rangle = -\langle Zy \mid Zx \rangle = \langle x \mid Z^2 y \rangle.$$

Being also compact, Z^2 induces an orthogonal eigendecomposition of V as

$$V = \bigoplus_{\mu \in \mathbb{R}} V(\mu)$$

where if $\mu \in \mathbb{R}$ then

$$V(\mu) = \{v \in V : Z^2 v = \mu v\}.$$

Note that if $\mu \in \mathbb{R}$ and $v \in V(\mu)$ then

$$\|Zv\|^2 = -\langle Z^2 v \mid v \rangle = -\mu \|v\|^2$$

whence $V(\mu) = 0$ when $\mu > 0$. Writing V^λ in place of $V(-\lambda^2)$ when λ is a non-negative real number, it follows that

$$V = V^0 \oplus \sum_{\lambda > 0} V^\lambda$$

where V^0 is the kernel of Z. Since Z is Hilbert-Schmidt, the positive λ for which $V^\lambda \neq 0$ form a square-summable (hence countable) set having 0 as the only possible limit point. Moreover, if $\lambda > 0$ and $V^\lambda \neq 0$ then V^λ is finite-dimensional: indeed, $\frac{1}{\lambda} Z \mid V^\lambda$ is a J-antilinear complex (hence *quaternionic*) structure, whence V^λ is even-dimensional and has an orthonormal (complex) basis consisting of pairs $v, \frac{1}{\lambda} Zv$ for suitable unit vectors $v \in V^\lambda$.

It is now plain that we may list the summable nonzero eigenvalues of $-Z^2$ by multiplicities as

$$\lambda_1^2, \lambda_1^2, \ldots, \lambda_j^2, \lambda_j^2, \ldots$$

and that $(\ker Z)^\perp = \Sigma_{\lambda>0} V^\lambda$ has a complete orthonormal system

$$x_1, y_1, \ldots, x_j, y_j, \ldots$$

with $Zx_j = \lambda_j y_j$ and $Zy_j = -\lambda_j x_j$ for $j > 0$. Fourier expansion now reveals that if $Z = Z_\zeta$ for $\zeta \in \bigwedge^2[V_J]$ as in Theorem 2.2.2 then

$$\zeta = \sum_{j>0} \lambda_j x_j \wedge y_j.$$

Our description of $\bigwedge^2[V_J]$ in terms of $\mathbb{S}(V_J)$ has thus produced a canonical form for its elements.

Theorem 2.2.3 *Let* $Z = Z_\zeta \in \mathbb{S}(V_J)$ *for* $\zeta \in \bigwedge^2[V_J]$. *There exist a square-summable positive sequence* $(\lambda_j : j > 0)$ *and a complete orthonormal sequence* $(x_j, y_j : j > 0)$ *for* $(\ker Z)^\perp$ *such that*

$$\zeta = \sum_{j>0} \lambda_j x_j \wedge y_j$$

with $Zx_j = \lambda_j y_j$ *and* $Zy_j = -\lambda_j x_j$ *for* $j > 0$. □

Thus armed, we are now able to establish that if $\zeta \in \bigwedge^2[V_J]$ and $n > 0$ then

$$\|\zeta^n\|^2 \leq n! \, \|\zeta\|^{2n}.$$

Retaining $Z = Z_\zeta \in \mathbb{S}(V_J)$ and the notation from above, let us introduce also $\zeta_j = x_j \wedge y_j$ for $j > 0$. Having even degree and being decomposable, the vectors $\{\zeta_j : j > 0\}$ in $\bigwedge^2[V_J]$ mutually commute and self-annihilate under wedge product. It follows that

$$\zeta^n = n! \sum_{j_1 < \cdots < j_n} \lambda_{j_1} \ldots \lambda_{j_n} \zeta_{j_1} \wedge \ldots \wedge \zeta_{j_n}$$

whence

$$\begin{aligned}
\|\zeta^n\|^2 &= (n!)^2 \sum \{\lambda_{j_1}^2 \ldots \lambda_{j_n}^2 : j_1, \ldots, j_n \text{ increasing}\} \\
&= n! \sum \{\lambda_{j_1}^2 \ldots \lambda_{j_n}^2 : j_1, \ldots, j_n \text{ distinct}\} \\
&\leq n! \sum \{\lambda_{j_1}^2 \ldots \lambda_{j_n}^2 : j_1, \ldots, j_n \text{ arbitrary}\} \\
&= n! \left(\sum_{j>0} \lambda_j^2\right)^n = n! \, \|\zeta\|^{2n}
\end{aligned}$$

as was promised.

Theorem 2.2.4 *If* $\zeta \in \bigwedge^2[V_J]$ *and if* $n > 0$ *then*
$$\|\zeta^n\|^2 \leq n! \, \|\zeta\|^{2n}.$$

□

As remarked when we announced this inequality, its importance for us is that it guarantees convergence in $\mathbb{H}_J(V)$ of the exponential series

$$\exp \zeta = \sum_{n \geq 0} \frac{1}{n!} \zeta^n.$$

Indeed, recalling that the spaces $\bigwedge^n[V_J]$ are orthogonal as n ranges over the non-negative integers, it follows from Theorem 2.2.4 that if $q > p$ then

$$\left\| \sum_{n=p+1}^{q} \frac{1}{n!} \zeta^n \right\|^2 = \sum_{n=p+1}^{q} \frac{1}{(n!)^2} \|\zeta^n\|^2$$

$$\leq \sum_{n=p+1}^{q} \frac{1}{n!} \|\zeta\|^{2n}$$

whence the indicated exponential series has Cauchy partial sums since the real exponential series $\sum_{n \geq 0} \|\zeta\|^{2n}/n!$ converges. Moreover, similar reasoning yields the following inequality.

Theorem 2.2.5   *If $\zeta \in \bigwedge^2[V_J]$ then the quadratic exponential*

$$\exp \zeta := \sum_{n \geq 0} \frac{1}{n!} \zeta^n \in \mathbb{H}_J(V)$$

satisfies the inequality

$$\| \exp \zeta \|^2 \leq \exp \|\zeta\|^2.$$

□

We refer to these *quadratic exponentials* also as *Gaussians*. They will play an important rôle in Section 3.3, when we determine precisely which Bogoliubov automorphisms are unitarily implementable in a Fock representation.

2.3 Creators and annihilators

Again, we choose and fix a unitary structure J on the real Hilbert space V and write $\langle \cdot \mid \cdot \rangle = \langle \cdot \mid \cdot \rangle_J$ for the resulting Hermitian inner product on the complex Hilbert space V_J. In the preceding section, we set up Fock space $\mathbb{H}_J(V)$: the complex Hilbert space that will carry the Fock representation induced by J. Here we develop the algebraic machinery of creators and annihilators: operators on Fock space from which we shall fashion the Fock representation in the following section. Initially, creators and annihilators are defined as operators on the exterior algebra $\bigwedge(V_J) = H_J(V)$ over V_J; then they are extended to Fock space $\mathbb{H}_J(V)$ by continuity, being bounded.

Creators are the simpler operators to define. If $v \in V$ then the *creator* (or *creation operator*) $c(v) = c_J(v)$ is the complex-linear operator on $H_J(V)$ given by

$$\zeta \in H_J(V) \quad \Rightarrow \quad c(v)\zeta = v \wedge \zeta.$$

Note that $c(Jv) = ic(v)$: thus, $c(v)$ is a complex-linear function of $v \in V_J$. Wedge product being anticommutative, it is at once clear that $c(v)^2 = 0$. Replacing v by $x + y$ it follows by polarization that

$$c(x)c(y) + c(y)c(x) = 0$$

whenever $x, y \in V$.

Annihilators are a little less simple to define. If $v \in V$ then the *annihilator* (or *annihilation operator*) $a(v) = a_J(v)$ is the complex-linear operator on $H_J(V)$ annihilating Ω_J and given on decomposables by

$$a(v)(w_0 \wedge \ldots \wedge w_n) = \sum_{j=0}^{n} (-1)^j \langle w_j \mid v \rangle \, w_0 \wedge \ldots \wedge \hat{w}_j \wedge \ldots \wedge w_n$$

for $w_0, \ldots, w_n \in V$; here, a circumflex signifies omission as usual. Somewhat more concisely, $a(v)$ is the unique complex-linear antiderivation of the exterior algebra $\bigwedge(V_J)$ such that $a(v)w = \langle w \mid v \rangle \Omega_J$ whenever $w \in V$: it satisfies

$$a(v)(\xi \wedge \eta) = (a(v)\xi) \wedge \eta + (-1)^p \, \xi \wedge (a(v)\eta)$$

for all $\xi \in \bigwedge^p(V_J)$ and $\eta \in \bigwedge^q(V_J)$. Note that $a(Jv) = -ia(v)$: thus, $a(v)$ is J-antilinear as a function of $v \in V_J$. As a consequence either of the fact that wedge product is anticommutative or of the next theorem, $a(v)^2 = 0$; by polarization, it follows that

$$a(x)a(y) + a(y)a(x) = 0$$

whenever $x, y \in V$.

Theorem 2.3.1 *If $v \in V$ and if $\xi, \eta \in H_J(V)$ then*

$$\langle c(v)\xi \mid \eta \rangle = \langle \xi \mid a(v)\eta \rangle.$$

Proof It is enough to check equality when ξ and η are decomposable; since the sum $H_J(V) = \bigoplus_{n \geq 0} \bigwedge^n(V_J)$ is orthogonal, we may suppose that $\xi = x_1 \wedge \ldots \wedge x_n$ and $\eta = y_0 \wedge y_1 \wedge \ldots \wedge y_n$ for vectors x_i and y_j in V. In this case,

$$\langle c(v)\xi \mid \eta \rangle = \det \begin{bmatrix} \langle v \mid y_0 \rangle & \cdots & \langle v \mid y_n \rangle \\ \vdots & & \vdots \\ \langle x_n \mid y_0 \rangle & \cdots & \langle x_n \mid y_n \rangle \end{bmatrix}.$$

This yields $\langle \xi \mid a(v)\eta \rangle$ precisely when expanded along the top row. $\qquad\square$

Thus, if $v \in V$ then the creator $c(v)$ and annihilator $a(v)$ are formally adjoint on $H_J(V)$. Creators and annihilators also satisfy the relations embodied in the next result.

Theorem 2.3.2　*If* $x, y \in V$ *then the identity*

$$c(x)a(y) + a(y)c(x) = \langle x \mid y \rangle I$$

holds among operators on $H_J(V)$.

Proof　By linearity, we need only apply both sides to a decomposable $\zeta \in \bigwedge^n(V_J)$. By factorization, we may assume $\zeta = \xi \wedge \eta$ for $\xi \in \bigwedge^p(V_J)$ and $\eta \in \bigwedge^q(V_J)$. Now, since $a(y)$ is an antiderivation we have

$$c(x)a(y)(\xi \wedge \eta) = c(x)\left[(a(y)\xi) \wedge \eta + (-1)^p \xi \wedge (a(y)\eta) \right]$$
$$= x \wedge (a(y)\xi) \wedge \eta + (-1)^p x \wedge \xi \wedge (a(y)\eta)$$

and

$$a(y)c(x)(\xi \wedge \eta) = a(y)(x \wedge \xi \wedge \eta)$$
$$= \langle x \mid y \rangle (\xi \wedge \eta)$$
$$- x \wedge (a(y)\xi) \wedge \eta - (-1)^p x \wedge \xi \wedge (a(y)\eta)$$

so

$$\left(c(x)a(y) + a(y)c(x) \right)(\xi \wedge \eta) = \langle x \mid y \rangle (\xi \wedge \eta)$$

completing the proof.　　　　　　　　　　　　　　　　　　□

We can now establish that creators and annihilators are bounded as operators on $H_J(V)$. Indeed, if $v \in V$ and $\zeta \in H_J(V)$ then

$$\|c(v)\zeta\|^2 + \|a(v)\zeta\|^2 = \langle c(v)\zeta \mid c(v)\zeta \rangle + \langle a(v)\zeta \mid a(v)\zeta \rangle$$
$$= \langle \left(a(v)c(v) + c(v)a(v) \right) \zeta \mid \zeta \rangle$$
$$= \|v\|^2 \|\zeta\|^2$$

so that $c(v)$ and $a(v)$ have operator norms at most $\|v\|$ on $H_J(V)$: in fact, they have operator norms equal to $\|v\|$ in view of the identities $c(v)\Omega_J = v$ and $a(v)v = \|v\|^2 \Omega_J$. We shall denote the continuous linear extensions of $c(v)$ and $a(v)$ to Fock space $\mathbb{H}_J(V)$ by the same symbols. Note that $c(v)$ and $a(v)$ are mutually adjoint operators on $\mathbb{H}_J(V)$ as a consequence of their being formally adjoint on $H_J(V)$ according to Theorem 2.3.1. We have more than verified the following.

Theorem 2.3.3　*If* $v \in V$ *then* $c(v)$ *and* $a(v)$ *extend continuously to mutually adjoint operators of norm* $\|v\|$ *on* $\mathbb{H}_J(V)$ *such that if* $\zeta \in \mathbb{H}_J(V)$ *then*

$$\|c(v)\zeta\|^2 + \|a(v)\zeta\|^2 = \|v\|^2 \|\zeta\|^2.　　　　　□$$

By continuity, the algebraic relationships between creators and annihilators established on $H_J(V)$ persist on $\mathbb{H}_J(V)$ as follows.

Theorem 2.3.4   *If $x, y \in V$ then on $\mathbb{H}_J(V)$ there hold the relations*
$$c(x)a(y) + a(y)c(x) = \langle x \mid y \rangle I$$
$$c(x)c(y) + c(y)c(x) = 0$$
$$a(x)a(y) + a(y)a(x) = 0.$$

□

The relations set forth in this theorem are the famous *canonical anticommutation relations* (CAR) satisfied by creators and annihilators on Fock space. These relations may be conveniently reformulated in terms of the *anticommutator bracket* $\{ , \}$ given by
$$\{X, Y\} = XY + YX$$
for elements X and Y in any algebra. In these terms, the canonical anticommutation relations assume the form
$$\{c(x), a(y)\} = \langle x \mid y \rangle I$$
$$\{c(x), c(y)\} = \{a(x), a(y)\} = 0.$$
We present a couple of easy consequences of the canonical anticommutation relations. First, recall that if $v \in V$ then $c(v)^2 = 0$ and $a(v)^2 = 0$ so that $\operatorname{ran} c(v) \subset \ker c(v)$ and $\operatorname{ran} a(v) \subset \ker a(v)$; in fact, these inclusions are equalities when v is nonzero. To see this, suppose for convenience that $v \in V$ is a unit vector and let $\zeta \in \ker c(v)$: then
$$\zeta = \{c(v), a(v)\} \zeta = c(v)\,(a(v)\zeta)$$
so that $\zeta \in \operatorname{ran} c(v)$; notice that the CAR pick out a specific preimage $a(v)\zeta$ of ζ under $c(v)$. Similarly, if $a(v)\zeta = 0$ then $\zeta = a(v)\,(c(v)\zeta)$.

Theorem 2.3.5   *Let $v \in V$ be a unit vector and let $\zeta \in \mathbb{H}_J(V)$. If $c(v)\zeta = 0$ then $\zeta = c(v)\,(a(v)\zeta)$ whilst if $a(v)\zeta = 0$ then $\zeta = a(v)\,(c(v)\zeta)$.*

□

Here, the restriction to unit vectors is of course inessential. In the second consequence of the CAR to which we draw attention, it is more important. We claim that if $v \in V$ is a unit vector then the creator $c(v)$ and annihilator $a(v)$ are *partial isometries*. By this, we mean that each is isometric on the orthocomplement of its kernel; this orthocomplement is called the initial space of the partial isometry, whose range is also called its final space. Now, consider the creator $c(v)$. From $c(v)^* = a(v)$

it follows that

$$(\ker c(v))^{\perp} = \overline{\operatorname{ran} a(v)}$$

$$= \operatorname{ran} a(v)$$

since $\operatorname{ran} a(v)$ is closed, being equal to $\ker a(v)$ by Theorem 2.3.5. If we let $\zeta \in \mathbb{H}_J(V)$ then

$$\|c(v)a(v)\zeta\| = \|a(v)\zeta\|$$

since

$$\langle c(v)a(v)\zeta \mid c(v)a(v)\zeta \rangle = \langle a(v)c(v)a(v)\zeta \mid a(v)\zeta \rangle$$

$$= \langle \{c(v), a(v)\}a(v)\zeta \mid a(v)\zeta \rangle$$

$$= \langle a(v)\zeta \mid a(v)\zeta \rangle$$

by virtue of $c(v)^* = a(v)$ and the CAR. Thus, $c(v)$ is indeed a partial isometry; its initial space is $\operatorname{ran} a(v) = \ker a(v)$ and its final space $\operatorname{ran} c(v) = \ker c(v)$. The annihilator $a(v)$ may be treated in a similar manner; alternatively, recall that the adjoint of a partial isometry is also a partial isometry, with initial space and final space interchanged. In formalizing these findings, we shall omit reference to final spaces since they are quite evident.

Theorem 2.3.6   *If $v \in V$ is a unit vector then $c(v)$ is a partial isometry with initial space $\operatorname{ran} a(v) = \ker a(v)$ and $a(v)$ is a partial isometry with initial space $\operatorname{ran} c(v) = \ker c(v)$.*   □

We note that this result may also be established using Theorem 2.3.3: if $v \in V$ is a unit vector and if $\zeta \in \mathbb{H}_J(V)$ then $\|c(v)\zeta\| = \|\zeta\|$ when and only when $a(v)\zeta = 0$. Of course, Theorem 2.3.3 itself is a consequence of the CAR.

Our dealings with the Fock representation require that we go further and determine precisely which Fock space vectors lie in the kernel of each creator and which lie in the kernel of each annihilator. Thus, the following result will yield for us irreducibility of the Fock representation.

Theorem 2.3.7

$$\bigcap_{v \in V} \ker a(v) = \mathbb{C}\Omega_J.$$

Proof One direction is plain enough: by their very definition, all annihilation operators annihilate Ω_J. Conversely, let $\zeta \in \mathbb{H}_J(V)$ be annihilated by $a(v)$ whenever $v \in V$. If $v_0, \ldots, v_m \in V$ then

$$\langle \zeta \mid v_0 \wedge \ldots \wedge v_m \rangle = \langle a(v_0)\zeta \mid v_1 \wedge \ldots \wedge v_m \rangle = 0$$

so that (by linearity) ζ is orthogonal to all elements of $\bigoplus_{n>0} \bigwedge^n(V_J)$ and hence lies in $\bigwedge^0(V_J) = \mathbb{C}\Omega_J$ as claimed. $\qquad\square$

Creators are somewhat more awkward to handle: in fact, the intersection of their kernels depends on whether the dimension of V is finite or infinite. Let $\{v_r : r \in \mathcal{R}\}$ be a complete orthonormal system for the complex Hilbert space V_J with $\mathcal{R}$ a totally ordered index set and recall that $\mathbb{H}_J(V)$ has $\{v_R : R \uparrow \mathcal{R}\}$ as a complete orthonormal system. Now, let $\zeta \in \mathbb{H}_J(V)$ lie in the kernel of every creator. Suppose first that V is infinite-dimensional. If $R = \{r_1, \ldots, r_n\}$ is a strictly increasing multiindex from $\mathcal{R}$ then choose $r \in \mathcal{R} - R$ and compute:

$$
\begin{aligned}
0 &= \langle c(v_r)\zeta \mid v_r \wedge v_R \rangle \\
  &= \langle \zeta \mid a(v_r)(v_r \wedge v_R) \rangle \\
  &= \langle \zeta \mid v_R \rangle
\end{aligned}
$$

since $a(v_r)$ is an antiderivation and the vectors $v_r, v_{r_1}, \ldots, v_{r_n}$ are orthonormal. The completeness of the orthonormal system $\{v_R : R \uparrow \mathcal{R}\}$ now forces $\zeta = 0$. Suppose instead that V_J has finite complex dimension m. The preceding argument shows that $\langle \zeta \mid v_R \rangle = 0$ whenever $R \uparrow \mathcal{R}$ has cardinality less than m; thus ζ lies in the top alternating power $\bigwedge^m(V_J)$.

Theorem 2.3.8 *If V is infinite-dimensional then $\bigcap_{v \in V} \ker c(v) = 0$ whilst if $\dim V_J = m$ then $\bigcap_{v \in V} \ker c(v) = \bigwedge^m(V_J)$.* $\qquad\square$

This result admits a simple refinement, which we shall need in our analysis of unitary implementability for Bogoliubov automorphisms in the Fock representation. Let X be a closed complex subspace of V_J and write $Y = X^\perp$ for its orthocomplement. Of course, $\bigwedge^n(Y) \subset \bigwedge^n(V_J)$ for each natural number n; moreover, the closure of $\bigoplus_{n \geq 0} \bigwedge^n(Y)$ in $\mathbb{H}_J(V)$ is canonically isomorphic to the Fock space $\mathbb{H}_J(Y)$ of the complex Hilbert space Y. We claim that the intersection $\bigcap_{x \in X} \ker c(x)$ is zero if X is infinite-dimensional and is $\bigwedge^m(X) \wedge \mathbb{H}_J(Y)$ if X is m-dimensional over $\mathbb{C}$. In preparation for our proof of this claim, let $\{x_s : s \in \mathcal{S}\}$ and $\{y_t : t \in \mathcal{T}\}$ be complete orthonormal systems for X and Y with totally ordered index sets $\mathcal{S}$ and $\mathcal{T}$ respectively; the vectors $x_S \wedge y_T$ then form a complete orthonormal system for $\mathbb{H}_J(V)$ when $S \uparrow \mathcal{S}$ and $T \uparrow \mathcal{T}$.

Now, let $\zeta \in \mathbb{H}_J(V)$ lie in the kernel of $c(x)$ for every $x \in X$. Assume first that X is infinite-dimensional. If $S \uparrow \mathcal{S}$ and $T \in \mathcal{T}$ then choose

$s \in \mathcal{S} - S$ and note that

$$0 = \langle c(x_s)\zeta \mid x_s \wedge x_S \wedge y_T \rangle$$
$$= \langle \zeta \mid a(x_s)(x_s \wedge x_S \wedge y_T)) \rangle$$
$$= \langle \zeta \mid x_S \wedge y_T \rangle.$$

Completeness of the orthonormal system $\{x_S \wedge y_T : S \uparrow \mathcal{S}, T \uparrow \mathcal{T}\}$ now forces $\zeta = 0$. Assume instead that X has finite complex dimension m. The argument just put forward establishes that $\langle \zeta \mid x_S \wedge y_T \rangle = 0$ if $S \uparrow \mathcal{S}$ has cardinality less than m and if $T \uparrow \mathcal{T}$. A glance at the Fourier expansion

$$\zeta = \sum_{S \uparrow \mathcal{S}, T \uparrow \mathcal{T}} \langle \zeta \mid x_S \wedge y_T \rangle x_S \wedge y_T$$

now reveals that ζ lies in the wedge product of $\bigwedge^m(X)$ and $\mathbb{H}_J(Y)$ as claimed.

Theorem 2.3.9   *Let X be a closed complex subspace of V_J. If X is infinite-dimensional then*

$$\bigcap_{x \in X} \ker c(x) = 0$$

whilst if $\dim X = m$ *then*

$$\bigcap_{x \in X} \ker c(x) = \bigwedge^m(X) \wedge \mathbb{H}_J(X^\perp).$$

                                                □

Our analysis of the Fock implementation of Bogoliubov automorphisms also calls for us to make explicit the action of annihilators on Gaussians. Thus, let $\zeta \in \bigwedge^2[V_J]$ correspond to the Hilbert-Schmidt antiskew operator $Z = Z_\zeta \in \mathbb{S}(V_J)$ given by

$$x, y \in V \quad \Rightarrow \quad \langle Zx \mid y \rangle = \langle \zeta \mid x \wedge y \rangle$$

as usual. If $v \in V$ then $a(v)\zeta = Zv$ since

$$\langle a(v)\zeta \mid u \rangle = \langle \zeta \mid v \wedge u \rangle = \langle Zv \mid u \rangle$$

for all $u \in V$. Induction now yields that if n is any positive integer then

$$a(v)(\zeta^n) = n(Zv) \wedge (\zeta^{n-1})$$

since $a(v)$ is an antiderivation. Lastly, by continuity it follows that the Gaussian

$$\exp \zeta = \sum_{n \geq 0} \frac{1}{n!} \zeta^n$$

satisfies

$$a(v)(\exp \zeta) = (Zv) \wedge (\exp \zeta)$$

and we have the following result.

Theorem 2.3.10 *If $\zeta \in \bigwedge^2[V_J]$ and if $v \in V$ then*

$$a(v)\zeta = Z_\zeta(v)$$

and

$$a(v)(\exp\zeta) = c(Z_\zeta v)(\exp\zeta).$$

$\square$

2.4 Fock representations

Having made all the necessary preparations, we at last have the where-withal to construct and examine the Fock representation of the C^* Clifford algebra $C[V]$ determined by the specific choice J of unitary structure on the real Hilbert space V.

This Fock representation will take place on the Fock space $\mathbb{H}_J = \mathbb{H}_J(V)$ and will be defined in terms of creators and annihilators, which will be denoted either by c_J and a_J for clarity or by c and a for convenience. For $v \in V$ we define a bounded linear operator $\pi_J(v)$ on $\mathbb{H}_J(V)$ by

$$\pi_J(v) = c_J(v) + a_J(v).$$

Note that if $v \in V$ then $\pi_J(v)$ is self-adjoint, since $c(v)$ and $a(v)$ are mutual adjoints as in Theorem 2.3.3. Note also from the canonical anticommutation relations in Theorem 2.3.4 that if $v \in V$ then $\pi_J(v)^2 = \|v\|^2 I$. Thus, the real-linear map π_J from V to $B(\mathbb{H}_J)$ is actually a self-adjoint Clifford map. The universal mapping property set forth in Theorem 1.2.4 now ensures that π_J extends to an isometric representation

$$\pi_J : C[V] \to B(\mathbb{H}_J)$$

of the C^* Clifford algebra over V. This is the *Fock representation* π_J of $C[V]$ determined by the choice of unitary structure J on V.

It is useful to note that the creators and annihilators from which the Fock representation π_J is fashioned can in fact be recovered as follows.

Theorem 2.4.1 *If $v \in V$ then*

$$c_J(v) = \tfrac{1}{2}\pi_J(v - \mathrm{i}Jv)$$

$$a_J(v) = \tfrac{1}{2}\pi_J(v + \mathrm{i}Jv).$$

Proof A straightforward computation. The v-dependence of $c(v)$ being

J-linear and that of $a(v)$ being J-antilinear, we see that

$$\pi_J(Jv) = c(Jv) + a(Jv)$$
$$= ic(v) - ia(v)$$

whence the result follows at once.   □

In particular, notice that all creators and annihilators lie in the range of the Fock representation. This has the immediate consequence (soon to be superceded) that the standard unit vector Ω_J is *cyclic* for π_J in the sense that $\{\pi_J(a)\Omega_J : a \in C[V]\}$ is dense in $\mathbb{H}_J$: indeed, the dense subspace H_J of $\mathbb{H}_J$ is spanned by decomposables, which may be obtained from Ω_J by successive application of creators. More explicitly, if $v_1, \ldots, v_n$ are orthogonal vectors in V_J then

$$\pi_J(v_1 \ldots v_n)\Omega_J = v_1 \wedge \ldots \wedge v_n$$

as follows easily from the canonical anticommutation relations. Incidentally, if $\{v_r : r \in \mathcal{R}\}$ is a complete orthonormal system for V_J and if $R = \{r_1 < \cdots < r_n\}$ is a multiindex from $\mathcal{R}$ then

$$\pi_J(v_R)\Omega_J = v_R$$

where v_R is a Clifford product on the left and wedge product on the right; our duplicated product notations are thus reconciled to some extent.

As another consequence of the fact that all creators and annihilators lie in the range of the Fock representation, π_J is *irreducible*: the bounded linear operators on $\mathbb{H}_J$ commuting with the range of π_J are precisely the scalars.

Theorem 2.4.2   *The Fock representation π_J is irreducible.*

Proof As recalled prior to the statement of the theorem, we must show that if $T \in B(\mathbb{H}_J)$ commutes with $\pi_J(v)$ whenever $v \in V$ then T is a scalar operator. To do this, note first that if $v \in V$ then the annihilator $a(v) = \frac{1}{2}\pi_J(v + iJv)$ commutes with T so that

$$a(v)T\,\Omega_J = Ta(v)\Omega_J = 0$$

and therefore $T\,\Omega_J = \mu\,\Omega_J$ for some scalar $\mu \in \mathbb{C}$ by Theorem 2.3.7. Now, if $a \in C[V]$ then

$$T\,\pi_J(a)\Omega_J = \pi_J(a)T\,\Omega_J = \mu\,\pi_J(a)\Omega_J$$

whence the fact that Ω_J is cyclic for π_J forces upon us the conclusion that $T = \mu I$. So ends the proof.   □

As a result, every nonzero vector in $\mathbb{H}_J(V)$ is cyclic for π_J.

We refer to the state of the C^* Clifford algebra $C[V]$ associated to the

Fock representation π_J by the cyclic unit vector $\Omega_J \in \mathbb{H}_J$ as the *Fock state* σ_J: thus, $\sigma_J : C[V] \to \mathbb{C}$ is given by the formula

$$a \in C[V] \quad \Rightarrow \quad \sigma_J(a) = \langle \pi_J(a)\Omega_J \mid \Omega_J \rangle_J$$

where we have recalled the label J on the Fock space inner product for emphasis. Notice in particular that if $x, y \in V$ then

$$\begin{aligned} \sigma_J(yx) &= \langle \pi_J(yx)\Omega_J \mid \Omega_J \rangle_J \\ &= \langle \pi_J(x)\Omega_J \mid \pi_J(y)\Omega_J \rangle_J \\ &= \langle x \mid y \rangle_J \end{aligned}$$

since if $v \in V$ then $v \in C[V]$ is self-adjoint, $c(v)\Omega_J = v$ and $a(v)\Omega_J = 0$. The observation

$$x, y \in V \quad \Rightarrow \quad \sigma_J(yx) = \langle x \mid y \rangle_J$$

can be generalized. Recall first that the complexification $V^{\mathbb{C}}$ carries a natural inner product $\langle \cdot \mid \cdot \rangle$; recall also (see Theorem 2.1.4 and Theorem 2.1.6) that P_J is the orthogonal projection from $V^{\mathbb{C}}$ onto the i-eigenspace F_J of the complexified operator J.

Theorem 2.4.3 *If $x, y \in V^{\mathbb{C}}$ then*

$$\sigma_J(y^*x) = \langle P_J x \mid P_J y \rangle.$$

Proof If $x = x_1 + \mathrm{i}x_2$ and $y = y_1 + \mathrm{i}y_2$ then both sides of the alleged equation reduce to

$$\langle x_1 \mid y_1 \rangle_J + \langle x_2 \mid y_2 \rangle_J + \mathrm{i}\langle x_2 \mid y_1 \rangle_J - \mathrm{i}\langle x_1 \mid y_2 \rangle_J :$$

the left side by reason of the special case presented prior to the theorem; the right side by definition of the natural inner product on $V^{\mathbb{C}}$. $\square$

We offer two specializations of this theorem. The first is that

$$v \in V \quad \Rightarrow \quad \sigma_J(v \cdot Jv) = \mathrm{i}\,\|v\|^2$$

which can also be seen as an instance of the special case preceding the theorem itself. The second is that

$$z \in \overline{F}_J \quad \Rightarrow \quad \sigma_J(z^*z) = 0$$

as follows from the circumstance that $\overline{F}_J = \ker P_J$.

In point of fact, these properties serve to distinguish the Fock state σ_J among all states of the C^* Clifford algebra: specifically, if σ is a state of $C[V]$ with the property that $\sigma(z^*z) = 0$ whenever $z \in \overline{F}_J$ then necessarily $\sigma = \sigma_J$. Our proof of this important fact will proceed in two stages: we begin by borrowing from physical tradition the notion of normal ordering, which leads to a decomposition of any given element in the complex Clifford algebra; we end by showing that the constant term in this decomposition is the value on the given element of any state satisfying the aforementioned property.

We require a couple of definitions. Let $w_1, \ldots, w_n$ be elements of $V^{\mathbb{C}} \subset C(V)$. We shall refer to the product $a = w_1 \ldots w_n$ as a *monomial* of degree n: we say that a is *polarized* if and only if each w_j lies in either F_J or $\overline{F}_J$; we say that a is in *normal form* if and only if it is polarized in such a way that all w_j in F_J lie to the left of all w_j in $\overline{F}_J$. By convention, monomials of degree zero are scalar multiples of the identity, which we consider to be polarized and in normal form.

Now, we claim that if again $w_1, \ldots, w_n \in V^{\mathbb{C}} \subset C(V)$ and if π is a permutation of the numbers $\{1, \ldots, n\}$ then the difference

$$w_1 \ldots w_n - (-1)^{\pi} w_{\pi(1)} \ldots w_{\pi(n)}$$

is a sum of monomials having degree at most $n - 2$. To see this, observe first that if π is merely a transposition of neighbouring indices then the claim follows at once from the fact that

$$x, y \in V^{\mathbb{C}} \quad \Rightarrow \quad xy + yx = 2(x \mid y)\mathbf{1}$$

where $(\cdot \mid \cdot)$ is extended to the canonical symmetric complex-bilinear form on $V^{\mathbb{C}}$. In the general case, π is a product of neighbour transpositions and the claim follows upon alternating summation. This claim furnishes the key to the result on decomposition of elements in the complex Clifford algebra.

Theorem 2.4.4 *Each element of $C(V)$ is a sum of monomials in normal form.*

Proof Let $w_1, \ldots, w_n \in V^{\mathbb{C}} \subset C(V)$; consider the monomial $w_1 \ldots w_n$. A multinomial expansion permits us to assume that $w_1 \ldots w_n$ is already polarized. Make the inductive hypothesis that any (polarized) monomial of degree less than n is a sum of monomials in normal form. Pick any permutation π of $\{1, \ldots, n\}$ such that $w_{\pi(1)} \ldots w_{\pi(n)}$ is in normal form. According to the claim established prior to the theorem, the difference

$$w_1 \ldots w_n - (-1)^{\pi} w_{\pi(1)} \ldots w_{\pi(n)}$$

is a sum of monomials having degree $n - 2$ or less; by induction, these are sums of monomials in normal form, whence so also is $w_1 \ldots w_n$ itself. The base step of our induction being quite transparent, we have shown that each monomial is a sum of monomials in normal form. It only remains to recall that each element of $C(V)$ is a sum of monomials.

$\square$

Thus, if $a \in C(V)$ then we may write $a = \mu\mathbf{1} + a'$ where $\mu \in \mathbb{C}$ and where a' is a sum of normal monomials having positive degree. Looking ahead a little, at the end of this section we shall determine the precise form of the constant term in this decomposition.

We are now equipped to demonstrate the advertised uniqueness of the Fock state. Thus, let $\sigma : C[V] \to \mathbb{C}$ be a state of the C^* Clifford algebra such that if $z \in \overline{F}_J$ then $\sigma(z^*z) = 0$. From the Cauchy-Schwarz inequality

$$a, b \in C[V] \quad \Rightarrow \quad |\sigma(b^*a)|^2 \le \sigma(a^*a)\,\sigma(b^*b)$$

it follows immediately that $\sigma(b^*a) = 0$ if either a or b lies in $\overline{F}_J$. In particular, σ vanishes on positive degree monomials in normal form. Now, if $a \in C(V)$ then Theorem 2.4.4 provides us with a decomposition $a = \mu\mathbf{1} + a'$ in which $\mu \in \mathbb{C}$ is a scalar and a' is a sum of positive degree monomials in normal form, whence $\sigma(a) = \mu$. This shows that the restriction of σ to $C(V)$ is uniquely determined; continuity forces uniqueness of σ over the whole of $C[V]$.

Theorem 2.4.5 *The Fock state σ_J is the unique state σ of $C[V]$ with the property that $\sigma(z^*z) = 0$ for all $z \in \overline{F}_J$.* $\quad\square$

This uniqueness property of the Fock state σ_J has some important consequences. First of all, it implies that the Fock state σ_J is *pure*: if

$$\sigma_J = \lambda'\sigma' + \lambda''\sigma''$$

for states σ', σ'' of $C[V]$ and positive reals λ', λ'' with sum unity, then in fact $\sigma' = \sigma'' = \sigma_J$. Indeed, if $z \in \overline{F}_J$ then

$$0 = \sigma_J(z^*z) = \lambda'\sigma'(z^*z) + \lambda''\sigma''(z^*z)$$

whence $\sigma'(z^*z) = \sigma''(z^*z) = 0$, as a result of which Theorem 2.4.5 forces upon us the desired conclusion $\sigma' = \sigma'' = \sigma_J$.

Theorem 2.4.6 *The Fock state σ_J is pure.* $\quad\square$

Of course, purity of the Fock state σ_J is equivalent to irreducibility of the Fock representation π_J according to the standard Gelfand-Naimark-Segal theory; we offered separate proofs for the sake of variety.

Another consequence of Theorem 2.4.5 has to do with Fock vacua. Let $\pi : C[V] \to B(\mathbb{H})$ be a representation of the C^* Clifford algebra. We shall refer to the nonzero vector $\Omega \in \mathbb{H}$ as a *J-vacuum vector* for π if and only if it satisfies the J-vacuum condition

$$v \in V \quad \Rightarrow \quad \pi(v + \mathrm{i}Jv)\Omega = 0$$

or equivalently

$$z \in \overline{F}_J \quad \Rightarrow \quad \pi(z)\Omega = 0.$$

For example, $\Omega_J \in \mathbb{H}_J$ is a J-vacuum vector for the Fock representation π_J: if $v \in V$ then $\pi_J(v + \mathrm{i}Jv) = 2a_J(v)$ by Theorem 2.4.1 and Ω_J is (up to scalar multiples) the unique vector in the kernel of every annihilator by Theorem 2.3.7; alternatively, if $z \in \overline{F}_J$ then

$$\|\pi_J(z)\Omega_J\|^2 = \langle \pi_J(z^*z)\Omega_J \mid \Omega_J \rangle_J$$
$$= \sigma_J(z^*z) = 0$$

according to Theorem 2.4.5.

Now, let $\pi : C[V] \to B(\mathbb{H})$ be a representation with a cyclic unit J-vacuum vector Ω and let $\sigma : C[V] \to \mathbb{C}$ be the corresponding state so that $\sigma(a) = \langle \pi(a)\Omega \mid \Omega \rangle$ for all $a \in C[V]$. We contend that π is then unitarily equivalent to the Fock representation π_J in such a way that Ω corresponds to the *Fock vacuum* Ω_J. To see that this is so, observe that the J-vacuum condition satisfied by Ω implies that if $z \in \overline{F}_J$ then

$$\sigma(z^*z) = \langle \pi(z^*z)\Omega \mid \Omega \rangle = \|\pi(z)\Omega\|^2 = 0.$$

Uniqueness of the Fock state as expressed in Theorem 2.4.5 now tells us that $\sigma = \sigma_J$. Our contention may now be justified by an appeal to the mechanics of the Gelfand-Naimark-Segal construction; however, it is easy enough to present the details for completeness. If $a \in C(V)$ then

$$\|\pi(a)\Omega\|^2 = \sigma(a^*a) = \sigma_J(a^*a) = \|\pi_J(a)\Omega_J\|^2$$

so that since $\pi_J(C(V)) \cdot \Omega_J = H_J$ we may well-define an isometry $U : H_J \to \mathbb{H}$ by

$$a \in C(V) \quad \Rightarrow \quad U\big(\pi_J(a)\Omega_J\big) = \pi(a)\Omega.$$

The continuous extension of U to $\mathbb{H}_J$ will be denoted by the same symbol; it is a unitary isomorphism $U : \mathbb{H}_J \to \mathbb{H}$ since $\Omega \in \mathbb{H}$ is cyclic for π. If a and b lie in $C[V]$ then by continuity

$$U\pi_J(a)\pi_J(b)\Omega_J = U\pi_J(ab)\Omega_J$$
$$= \pi(ab)\Omega = \pi(a)\pi(b)\Omega$$
$$= \pi(a)U\pi_J(b)\Omega_J$$

so that

$$a \in C[V] \quad \Rightarrow \quad U\pi_J(a) = \pi(a)U$$

since $\Omega_J \in \mathbb{H}_J$ is cyclic for π_J. Finally, the unitary isomorphism $U : \mathbb{H}_J \to \mathbb{H}$ intertwining π_J and π plainly maps Ω_J to Ω and is unique by irreducibility of π_J.

Theorem 2.4.7 *If $\pi : C[V] \to B(\mathbb{H})$ is a representation with a cyclic* *unit J-vacuum vector Ω then there exists a unique unitary isomorphism $U : \mathbb{H}_J \to \mathbb{H}$ such that $U \cdot \Omega_J = \Omega$ and such that*

$$a \in C[V] \quad \Rightarrow \quad \pi(a) = U\pi_J(a)U^*.$$

$\square$

Notice in particular that the existence of a cyclic vacuum vector forces a representation of the C^* Clifford algebra to be irreducible.

Theorem 2.4.7 will be of considerable service to us in recognizing Fock representations: see our discussion of Fock implementable Bogoliubov automorphisms in Section 3.3; see also our discussion of inner Bogoliubov automorphisms for the vN Clifford algebra in Section 4.3. We repeat: in order to show that a given representation $\pi : C[V] \to B(\mathbb{H})$ is unitarily equivalent to π_J it is enough to find a cyclic J-vacuum vector $\Omega \in \mathbb{H}$ for π.

Our last task in this section is to deal with a matter raised after Theorem 2.4.4: to provide an alternative explicit formula for the Fock state σ_J arising from $J \in \mathbb{U}(V)$. Of course, σ_J is an even state: it vanishes on the odd part $C^-[V]$ of the C^* Clifford algebra. We claim that if $v_1, \ldots, v_{2n} \in V$ then

$$\sigma_J(v_1 \ldots v_{2n}) = (-1)^{\frac{1}{2}n(n-1)} \sum_p (-1)^p \prod_{i=1}^n \sigma_J(v_{p(i)}v_{p(n+i)})$$

where $(-1)^p$ is the sign of p as p runs over all permutations of $\{1, \ldots, 2n\}$ such that $p(1) < \cdots < p(n)$ and $p(i) < p(n+i)$ for $i \in \{1, \ldots, n\}$. Since $(-1)^{\frac{1}{2}n(n-1)}$ is the sign of the permutation

$$\begin{pmatrix} 1 & 2 & \ldots & 2i-1 & 2i & \ldots & 2n-1 & 2n \\ 1 & n+1 & \ldots & i & n+i & \ldots & n & 2n \end{pmatrix}$$

we equivalently claim that

$$\sigma_J(v_1 \ldots v_{2n}) = \sum (-1)^{(q,r)} \prod_{i=1}^n \sigma_J(v_{q_i} v_{r_i})$$

where $(-1)^{(q,r)}$ is the sign of (q,r) as (q,r) runs over the set $\mathcal{P}$ of all permutations

$$\begin{pmatrix} 1 & 2 & \ldots & 2n-1 & 2n \\ q_1 & r_1 & \ldots & q_n & r_n \end{pmatrix}$$

such that $q_1 < \ldots < q_n$ and $q_i < r_i$ for $i \in \{1, \ldots, n\}$. Our proof of this claim will be inductive, based on properties of permutations and on the canonical anticommutation relations. The base of the induction is transparent and requires no further comment.

In discussing permutations, the following agreement will be convenient. If $\mathcal{I} = \{i_1, \ldots, i_m\}$ is a set of positive integers in increasing order, then we shall write $(j_1 \bullet \cdots \bullet j_m)$ for the permutation

$$\begin{pmatrix} i_1 & \cdots & i_m \\ j_1 & \cdots & j_m \end{pmatrix}$$

of $\mathcal{I}$. With this agreement, the permutation (q, r) appearing in the formula claimed for σ_J would alternatively be written as $(q_1 \bullet r_1 \bullet \cdots \bullet q_n \bullet r_n)$.

Now, for $j \in \{1, \ldots, 2n-1\}$ let us denote by $\mathcal{P}(j)$ the collection of all permutations $(s, t) = (s_1 \bullet t_1 \bullet \cdots \bullet s_{n-1} \bullet t_{n-1})$ of $\{1, \ldots, 2n-1\} - \{j\}$ such that $s_1 < \cdots < s_{n-1}$ and $s_i < t_i$ whenever $i \in \{1, \ldots, n-1\}$. A canonical bijection between $\mathcal{P}$ and the union $\mathcal{P}(1) \cup \ldots \cup \mathcal{P}(2n-1)$ is defined as follows. In the one direction, let $(s, t) = (s_1 \bullet t_1 \bullet \cdots \bullet s_{n-1} \bullet t_{n-1}) \in \mathcal{P}(j)$ for $j \in \{1, \ldots, 2n-1\}$. With the obvious understanding at endpoints, if $i \in \{1, \ldots, n\}$ and $s_{i-1} < j < s_i$ then for $k \in \{1, \ldots, n\}$ we put

$$q_k = \begin{cases} s_k & (k < i) \\ j & (k = i) \\ s_{k-1} & (k > i) \end{cases}$$

and

$$r_k = \begin{cases} t_k & (k < i) \\ 2n & (k = i) \\ t_{k-1} & (k > i) \end{cases}.$$

Now map (s, t) to $\chi(s, t) = (q, r) = (q_1 \bullet r_1 \bullet \cdots \bullet q_n \bullet r_n) \in \mathcal{P}$. In the other direction, $\chi^{-1} : \mathcal{P} \to \mathcal{P}(1) \cup \cdots \cup \mathcal{P}(2n-1)$ is straightforward to specify explicitly.

Of crucial importance is the way in which the bijection χ just defined affects the sign of a permutation. We contend that if $(s, t) \in \mathcal{P}(j)$ for $j \in \{1, \ldots, 2n-1\}$ then

$$(-1)^{\chi(s,t)} = (-1)^{j-1}(-1)^{(s,t)}.$$

The formal definition of χ explains our choice to present the proof of this contention schematically. In terms of our established notation, the permutation

$$\chi(s,t) = \begin{pmatrix} 1 & 2 & \cdots & \cdots & \cdots & \cdots & \cdots & \cdots & 2n-1 & 2n \\ s_1 & t_1 & \cdots & t_{i-1} & j & 2n & s_i & \cdots & s_{n-1} & t_{n-1} \end{pmatrix}$$

has the same sign as

$$\begin{pmatrix} 1 & 2 & \cdots & \cdots & \cdots & \cdots & 2n-1 & 2n \\ j & s_1 & \cdots & t_{i-1} & s_i & \cdots & t_{n-1} & 2n \end{pmatrix}$$

which is $(-1)^{j-1}$ times the sign of

$$\begin{pmatrix} j & 1 & \cdots & j-1 & j+1 & \cdots & 2n-1 & 2n \\ j & s_1 & \cdots & \cdots & \cdots & \cdots & t_{n-1} & 2n \end{pmatrix}$$

which in turn is plainly the sign of (s, t) itself.

Theorem 2.4.8 *If $j \in \{1, \ldots, 2n - 1\}$ and $(s, t) \in \mathcal{P}(j)$ then*
$$(-1)^{\chi(s,t)} = (-1)^{j-1}(-1)^{(s,t)}.$$

□

Finally, let $v_1, \ldots, v_{2n} \in V$. Since $a(v_{2n})\Omega_J = 0$ it follows from the canonical anticommutation relations in Theorem 2.3.4 that

$$\pi_J(v_1 \ldots v_{2n})\Omega_J = \pi_J(v_1) \ldots \pi_J(v_{2n-1}) \, c(v_{2n})\Omega_J$$
$$= \sum_{j=1}^{2n-1} (-1)^{j-1}\langle v_{2n} \mid v_j\rangle \pi_J(v_1 \ldots \hat{v}_j \ldots v_{2n-1})\Omega_J$$
$$- c(v_{2n})\pi_J(v_1 \ldots v_{2n-1})\Omega_J$$

by passing $c(v_{2n})$ to the left of the operators $\pi_J(v_{2n-1}), \ldots, \pi_J(v_1)$ in turn; here, the circumflex indicates omission as usual. As

$$\langle c(v_{2n})\pi_J(v_1 \ldots v_{2n-1})\Omega_J \mid \Omega_J\rangle = \langle \pi_J(v_1 \ldots v_{2n-1})\Omega_J \mid a(v_{2n})\Omega_J\rangle = 0$$

it now follows upon taking inner products against Ω_J that

$$\sigma_J(v_1 \ldots v_{2n}) = \sum_{j=1}^{2n-1} (-1)^{j-1}\sigma_J(v_j \, v_{2n})\sigma_J(v_1 \ldots \hat{v}_j \ldots v_{2n-1})$$

by virtue of Theorem 2.4.3. Inductively,

$$\sigma_J(v_1 \ldots \hat{v}_j \ldots v_{2n-1}) = \sum_{(s,t)} (-1)^{(s,t)} \prod_{i=1}^{n-1} \sigma_J(v_{s_i} \, v_{t_i})$$

from which it follows by substitution that

$$\sigma_J(v_1 \ldots v_{2n}) = \sum_{(q,r)} (-1)^{(q,r)} \prod_{i=1}^{n} \sigma_J(v_{q_i} \, v_{r_i})$$

in established notation, on account of Theorem 2.4.8 concerning the bijection $\chi : \mathcal{P}(1) \cup \cdots \cup \mathcal{P}(2n - 1) \to \mathcal{P}$. The explicit formula claimed for σ_J is thus verified; we record its original form.

Theorem 2.4.9 *The Fock state σ_J is even and if $v_1, \ldots, v_{2n} \in V$ then*

$$\sigma_J(v_1 \ldots v_{2n}) = (-1)^{\frac{1}{2}n(n-1)} \sum_{p} (-1)^p \prod_{i=1}^{n} \sigma_J(v_{p(i)} \, v_{p(n+i)})$$

where p runs over all permutations of $\{1, \ldots, 2n\}$ satisfying $p(1) < \cdots < p(n)$ and $p(i) < p(n + i)$ whenever $i \in \{1, \ldots, n\}$. □

2.5 Parity considerations

In this, the closing section of the present chapter, we fix once again a unitary structure J on the real Hilbert space V and study the restriction of the corresponding Fock representation $\pi_J : C[V] \to B(\mathbb{H}_J)$ to the even C^* Clifford algebra $C^+[V]$. It turns out that this restriction is no longer irreducible: rather, it decomposes as the orthogonal sum of two inequivalent irreducible subrepresentations; in our model, these act on the subspaces of Fock space $\mathbb{H}_J$ formed by closing the sum of the spaces $\bigwedge^n(V_J)$ over n even on the one hand and over n odd on the other.

Explicitly, let us define closed subspaces $\mathbb{H}_J^+$ and $\mathbb{H}_J^-$ of Fock space $\mathbb{H}_J$ by

$$\mathbb{H}_J^+ = \left\{ \bigoplus_{m \geq 0} {\bigwedge}^{2m}(V_J) \right\}^-$$

and

$$\mathbb{H}_J^- = \left\{ \bigoplus_{m \geq 0} {\bigwedge}^{2m+1}(V_J) \right\}^-$$

where an upper bar on the right indicates closure. Let us also introduce a bounded linear operator Γ_J on $\mathbb{H}_J$ by demanding that

$$n \geq 0 \quad \Rightarrow \quad \Gamma_J \mid {\bigwedge}^n(V_J) = (-I)^n.$$

Thus

$$\Gamma_J \mid \mathbb{H}_J^\pm = \pm I$$

and

$$\mathbb{H}_J^\pm = (I \pm \Gamma_J)\mathbb{H}_J.$$

Plainly, the closed subspaces $\mathbb{H}_J^+$ and $\mathbb{H}_J^-$ are orthocomplementary. Also, the operator Γ_J is a symmetry, being both self-adjoint and unitary.

Recall that if $v \in V$ and $n \in \mathbb{N}$ then the creator $c(v)$ maps $\bigwedge^n(V_J)$ to $\bigwedge^{n+1}(V_J)$ and the annihilator $a(v)$ maps $\bigwedge^n(V_J)$ to $\bigwedge^{n-1}(V_J)$ where $\bigwedge^{-1}(V_J) := \{0\}$. From this recollection, it follows that if $v \in V$ then $\pi_J(v)$ maps each of the spaces $\mathbb{H}_J^\pm$ to the other and anticommutes with Γ_J so that

$$\Gamma_J \, \pi_J(v)\Gamma_J = -\pi_J(v).$$

More generally, this in turn implies that Γ_J implements the grading automorphism $\gamma = \theta_{-I}$ of $C[V]$ in the Fock representation π_J as follows.

Theorem 2.5.1 *If $a \in C[V]$ then*

$$\pi_J(\gamma a) = \Gamma_J \, \pi_J(a)\Gamma_J. \qquad \square$$

For this reason, we shall call Γ_J the *grading operator* on Fock space $\mathbb{H}_J$.

Accordingly, we shall call $\mathbb{H}_J^+$ the *even Fock space* (referring to its elements as even) and call $\mathbb{H}_J^-$ the *odd Fock space* (referring to its elements as odd).

We draw from Theorem 2.5.1 the conclusion that if a lies in the even C^* Clifford algebra $C^+[V]$ then $\pi_J(a)$ commutes with the grading operator Γ_J and so leaves each of the subspaces $\mathbb{H}_J^+$ and $\mathbb{H}_J^-$ invariant. In this way, we obtain upon restriction the *even Fock representation*

$$\pi_J^+ : C^+[V] \to B(\mathbb{H}_J^+)$$

and the *odd Fock representation*

$$\pi_J^- : C^+[V] \to B(\mathbb{H}_J^-).$$

It is upon these representations of $C^+[V]$ that we focus in the remainder of this section.

The representations π_J^+ and π_J^- are in fact irreducible. In order to prove this, it is convenient first to identify the von Neumann algebra $\pi_J(C^+[V])''$ on $\mathbb{H}_J$ generated by the image of $C^+[V]$ under π_J. Before doing so, observe that

$$\pi_J(C[V])'' = B(\mathbb{H}_J)$$

since π_J is irreducible.

Theorem 2.5.2 $\pi_J(C^+[V])'' = \{\Gamma_J\}'.$

Proof Theorem 2.5.1 already implies that $\Gamma_J \in \pi_J(C^+[V])'$ whence $\pi_J(C^+[V])'' \subset \{\Gamma_J\}'$. For the reverse inclusion, let $T \in \{\Gamma_J\}'$ and choose a net $(a_j : j \in \mathcal{J})$ in $C[V]$ such that $\pi_J(a_j) \xrightarrow{w} T$. Since Γ_J implements γ and commutes with T we also have $\pi_J(\gamma a_j) \xrightarrow{w} T$. Upon averaging, it follows that T is the weak operator limit of the net $\left(\pi_J\left(\frac{1}{2}(a_j + \gamma a_j)\right) : j \in \mathcal{J} \right)$ in $\pi_J(C^+[V])$ and hence lies in $\pi_J(C^+[V])''$. This concludes the proof. $\square$

Now the representations π_J^+ and π_J^- are irreducible, as announced: we can handle both together, as follows. Let $T^+ \in B(\mathbb{H}_J^+)$ commute with $\pi_J^+(a)$ and let $T^- \in B(\mathbb{H}_J^-)$ commute with $\pi_J^-(a)$ for all $a \in C^+[V]$. The operator $T \in B(\mathbb{H}_J)$ determined by the requirement that $T \mid \mathbb{H}_J^\pm = T^\pm$ then commutes with $\pi_J(a)$ whenever $a \in C^+[V]$ and so lies in $\pi_J(C^+[V])' = \{\Gamma_J\}''$ by Theorem 2.5.2. As a consequence, each of $T^\pm = T \mid \mathbb{H}_J^\pm$ is a scalar operator. Each of the commutants $\pi_J^+(C^+[V])' \subset B(\mathbb{H}_J^+)$ and $\pi_J^-(C^+[V])' \subset B(\mathbb{H}_J^-)$ being thus scalar, it follows that each of the representations π_J^+ and π_J^- is indeed irreducible.

Theorem 2.5.3 *The representations π_J^+ and π_J^- of the even C^* Clifford algebra $C^+[V]$ are irreducible.* □

In addition to being irreducible, the representations π_J^+ and π_J^- are also inequivalent. A suspicion that this is the case stems from the circumstance that the Fock vacuum Ω_J lies in $\mathbb{H}_J^+$ and not in $\mathbb{H}_J^-$. However, the J-vacuum condition on Ω_J is formulated in terms of the Fock representation of $C[V]$ rather than in terms of its restriction to $C^+[V]$. The following result remedies this situation, once it is noticed that if $v \in V$ then the operator $c(v)a(v)$ lies in $\pi_J(C^+[V])$.

Theorem 2.5.4 *Up to scalar multiples, $\Omega_J \in \mathbb{H}_J$ is the unique vector in the kernel of $c(v)a(v)$ for each $v \in V$.*

Proof One direction is of course plain. For the other, let $\zeta \in \mathbb{H}_J$ and suppose that $c(v)a(v)\zeta = 0$ whenever $v \in V$. If indeed $v \in V$ then
$$\|a(v)\zeta\|^2 = \langle a(v)\zeta \mid a(v)\zeta \rangle = \langle c(v)a(v)\zeta \mid \zeta \rangle = 0$$
whence $a(v)\zeta = 0$ and therefore $\zeta \in \mathbb{C}\Omega_J$ by Theorem 2.3.7. □

Inequivalence of the representations π_J^+ and π_J^- is now almost a triviality. Suppose $T : \mathbb{H}_J^+ \to \mathbb{H}_J^-$ to be a unitary isomorphism such that $\pi_J^-(a)T = T\,\pi_J^+(a)$ whenever $a \in C^+[V]$. If $v \in V$ then
$$c(v)a(v) = \tfrac{1}{2}\pi_J\big(\|v\|^2 \mathbf{1} + iv \cdot Jv\big) \in \pi_J\big(C^+[V]\big)$$
thus
$$c(v)a(v)T\,\Omega_J = T\,c(v)\,a(v)\,\Omega_J = 0$$
and so $T\,\Omega_J \in \mathbb{C}\Omega_J$ by Theorem 2.5.4. This is absurd, since $\Omega_J \in \mathbb{H}_J^+$ whereas $T\,\Omega_J \in \mathbb{H}_J^-$.

Theorem 2.5.5 *The representations π_J^+ and π_J^- of the even C^* Clifford algebra $C^+[V]$ are inequivalent.* □

An alternative proof of this theorem may be based on Theorem 2.5.2 and runs as follows. Suppose again that $T : \mathbb{H}_J^+ \to \mathbb{H}_J^-$ is a unitary isomorphism with $\pi_J^-(a)T = T\,\pi_J^+(a)$ whenever $a \in C^+[V]$. Define a unitary operator (indeed, a symmetry) U on $\mathbb{H}_J$ by
$$U = \begin{bmatrix} 0 & T^* \\ T & 0 \end{bmatrix}$$
in block form relative to the orthogonal decomposition $\mathbb{H}_J = \mathbb{H}_J^+ \oplus \mathbb{H}_J^-$. The supposed intertwining property of T entails that U commutes with $\pi_J(a)$ whenever $a \in C^+[V]$ and therefore that
$$U \in \pi_J(C^+[V])' = \{\Gamma_J\}''.$$

It follows that U acts on each of $\mathbb{H}_J^+$ and $\mathbb{H}_J^-$ as a scalar, contrary to construction. This contradiction proves once again that the representations π_J^+ and π_J^- are inequivalent.

All of the foregoing has been formulated in terms of our specific model of the Fock representation. More abstractly, suppose that we are given a representation $\pi : C[V] \to B(\mathbb{H})$ with $\Omega \in \mathbb{H}$ a cyclic unit J-vacuum vector. Theorem 2.4.7 informs us that π is uniquely unitarily equivalent to the Fock representation π_J in such a way that Ω corresponds to the Fock vacuum Ω_J. Of course, the grading operator Γ_J on $\mathbb{H}_J$ thereby induces a grading operator Γ on $\mathbb{H}$ whose eigenspaces $\mathbb{H}^\pm = (I \pm \Gamma)\mathbb{H}$ support representations π^+ and π^- of the even C^* Clifford algebra $C^+[V]$. Intrinsically, the symmetry Γ is the unique unitary operator on $\mathbb{H}$ fixing Ω and implementing γ in π; alternatively, $\mathbb{H}^+$ is the completion of the subspace of $\mathbb{H}$ generated from Ω by application of the image of $C^+[V]$ under π.

Remarks

Unitary structures

The constructions in this chapter rest upon providing the (other than odd-dimensional) real Hilbert space V with a unitary structure. As we have seen in the abstract, such always exist and indeed constitute a homogeneous space for the orthogonal group on V. It is important to realize that unitary structures arise naturally, primarily in the quantum theory of fermions. For accounts of this material, we refer to [15] on quantum statistical mechanics and to [86] on the Dirac equation. We also refer to the recent text [7] on quantum field theory; this incorporates information on the fermionic quantization of orthogonal dynamics, describing a natural procedure for selecting a unitary structure and the corresponding Fock representation.

Spin representations

Let $J \in \mathbb{U}(V)$ be a unitary structure on the real Hilbert space V. We have chosen to conform with historical (physical) tradition in referring to π_J as a Fock representation of the C^* Clifford algebra $C[V]$. We could instead have referred to π_J as a spin representation of $C[V]$ and to elements of $\mathbb{H}_J$ as spinors; in these terms, π_J^+ and π_J^- would be half-spin representations of the even C^* Clifford algebra. We make one special observation regarding the Fock (or spin) representation π_J when V is even-dimensional. In this case, a simple count reveals that $C(V)$ and the operator algebra $B(\mathbb{H}_J)$ have equal complex dimension; the faithful representation π_J is therefore actually an isomorphism. As a consequence of this, we see that $C(V)$ is isomorphic to a full complex matrix algebra. Of course, it follows that all irreducible representations of $C(V)$ are equivalent when V is even-dimensional. The corresponding assertion for infinite-dimensional V fails in quite spectacular fashion, as we shall see in the next chapter.

Holomorphic spinors

A unitary structure J on the real Hilbert space V gives rise to the Fock representation π_J by other means than the exterior algebra construction presented in the text. Special mention must be made of the holomorphic spinor construction due to Shale & Stinespring, appearing first in [80]. The idea of this construction is as follows. Let $\mathbb{H}_\tau$ be the complex Hilbert space obtained by completing $C(V)$ in the inner product determined by its canonical trace; let $\Omega := \mathbf{1} \in C(V) \subset \mathbb{H}_\tau$ be its standard unit vector and let Γ be the unitary operator on $\mathbb{H}_\tau$ that restricts to $C(V)$ as the grading automorphism. If for $v \in V$ and $\zeta \in \mathbb{H}_\tau$ we put

$$\pi'_J(v)\zeta = \tfrac{1}{\sqrt{2}}\{v \cdot \zeta - \mathrm{i}\Gamma(\zeta) \cdot Jv\}$$

then $\pi'_J : V \to B(\mathbb{H}_\tau)$ is a self-adjoint Clifford map and so extends to a star-representation π'_J of $C[V]$ on $\mathbb{H}_\tau$. As it happens, this representation π'_J itself is not irreducible. However, its cyclic subrepresentation π''_J generated by the standard unit vector Ω is irreducible. Indeed, Theorem 2.4.7 implies that π''_J and π_J are unitarily equivalent, since Ω is evidently a (cyclic) J-vacuum vector for π''_J. Modulo sign conventions, π''_J is the holomorphic spinor version of π_J. See [80] for more information on the holomorphic spinor representation, including an explanation of the name.

Quasifree states

Again, let V be an infinite-dimensional real Hilbert space. It is not difficult to show that to each state σ of $C[V]$ there corresponds a skew-adjoint operator C of norm at most unity on V such that

$$x, y \in V \quad \Rightarrow \quad \sigma(yx) = (x \mid y) + \mathrm{i}(x \mid Cy);$$

we refer to C as the *covariance* of σ. For example, if $J \in \mathbb{U}(V)$ then the Fock state σ_J has J as its covariance; also, the canonical trace τ has covariance zero. In general, a state of $C[V]$ is not completely specified by its covariance alone. However covariance does specify a state of $C[V]$

when the state has the special property of being quasifree. We say that the state σ of $C[V]$ is *quasifree* if and only if it vanishes on $C^-[V]$ and satisfies the condition that if $v_1, \ldots, v_{2n} \in V$ then

$$\sigma(v_1 \ldots v_{2n}) = (-1)^{\binom{n}{2}} \sum_p (-1)^p \prod_{j=1}^n \sigma\left(v_{p(j)}\, v_{p(n+j)}\right)$$

where summation extends over all permutations p of $\{1, \ldots, 2n\}$ with $p(1) < \ldots < p(n)$ and $p(j) < p(n+j)$ for $j \in \{1, \ldots, n\}$ and where $(-1)^p$ is the sign of p. In reverse, the formulae above in fact define a state σ of $C[V]$ with C as covariance. Thus, there is a natural bijective correspondence between quasifree states and covariance operators. For example, Fock states of $C[V]$ are quasifree, as we saw at the end of Section 2.4; in fact, Fock states are distinguished among all quasifree states by their purity. The canonical trace τ is also quasifree; of course, it is distinguished by being central. Lest it be imagined that all states of $C[V]$ are quasifree, see [40]. Information on quasifree states may be found by turning to [2] [8] [55] [56] [72] for example.

History and miscellany

The canonical anticommutation relations were introduced in 1928 by Jordan & Wigner [49] in connection with quantizing the electron field; here it is shown that the CAR over a finite-dimensional complex Hilbert space generate a full matrix algebra. The Fock space formalism of course originated with Fock himself [36] in 1932. It was placed on a firm mathematical foundation by Cook [28] following work presented to the National Academy of Sciences in 1951. The Fock (or Fock-Cook) representation served as the principal model for fermionic (second) quantization until later in the 1950s, when its inadequacies began to surface: see the historical Remarks at the end of the next chapter. Notwithstanding the ravages of progress, the Fock representation of course remains the fundamental model describing a free fermion field. For additional information, we direct attention to [7] [15] [86].

Now for some more specific comments. It is usual to speak of $J \in \mathbb{U}(V)$ as a complex structure, leaving implicit the assumption that it be compatible with the underlying real inner product. We have chosen the more informative name of unitary structure since J converts V into a unitary space. Regarding terminology for the alternatives, F_J is a *polarization* while Araki [3] refers to P_J as a *basis projection*. A discussion of Fock spaces (as exterior hilbertian powers) appears in Chapter 5 of [14]; see also the exercises in Chapter 12 of [50]. Our account of uniqueness of the Fock state σ_J determined by $J \in \mathbb{U}(V)$ is in essence taken from [3] by Araki, as is our alternative proof that π_J^+ and π_J^- are inequivalent. Finally, there exist alternative conventions regarding wedge product and inner product on Fock space; these lead to scaling factors in the definition of creators and annihilators.

3

INTERTWINING OPERATORS

In this third chapter, we examine the effect of varying unitary structures on the Fock representations to which they give rise. On the one hand, we consider the *equivalence problem*: if J and K are unitary structures on V then we ask for necessary and sufficient conditions in order that the Fock representations π_J and π_K be unitarily equivalent, in the sense that there exists a unitary isomorphism $T : \mathbb{H}_J \to \mathbb{H}_K$ with the property that $\pi_K(a) = T \, \pi_J(a) \, T^*$ whenever $a \in C[V]$. The solution to the equivalence problem asserts that π_J and π_K are unitarily equivalent if and only if the difference $K - J$ is a Hilbert-Schmidt operator on V. This leads directly to a proof of the fact that if V is infinite-dimensional then $C[V]$ has uncountably many unitarily inequivalent Fock representations. On the other hand, we consider the *implementation problem*: if J is a fixed unitary structure on V then we ask for necessary and sufficient conditions on the orthogonal transformation $g \in O(V)$ in order that the Bogoliubov automorphism θ_g of $C[V]$ be implemented in the Fock representation π_J by a unitary operator U on $\mathbb{H}_J$ such that $\pi_J(\theta_g a) = U\pi_J(a)U^*$ whenever $a \in C[V]$. The solution to the implementation problem asserts that θ_g is unitarily implemented in π_J if and only if the commutator $gJ - Jg$ is a Hilbert-Schmidt operator on V. This accounts for the significance of the famous restricted orthogonal group, usually denoted $O_{\mathrm{res}}(V)$ but here denoted $O_J(V)$ for clarity. Note that if V is even-dimensional then $C[V] = C(V)$ is isomorphic to a full complex matrix algebra. As a consequence, all of its irreducible representations are equivalent; in addition, each of its automorphisms is inner and hence implemented in any representation.

We begin in §1 by disposing of some preparatory material relating to decompositions of orthogonal transformations induced by a choice of unitary structure. In §2 we demonstrate the intimate relationship between the problems of equivalence and implementation; we also consider some special cases. Our solution to the implementation problem is presented in §3; our solution to the equivalence problem is deduced in §4. In both of these sections we develop a number of consequences that follow from our solutions. Lastly, §5 is devoted to the consideration of parity issues. Among other things, the concluding Remarks offer references to additional information concerning the restricted orthogonal group and its group of unitary implementers.

3.1 Orthogonal transformations

Before broaching the principal topic of this chapter, we prepare the way by dissecting orthogonal transformations. Fixing a choice J of unitary structure for the real Hilbert space V as usual, there result two related dissections of any given orthogonal transformation $g \in O(V)$: on the one hand, g admits an additive decomposition as $g = C_g + A_g$ in which C_g is J-linear and A_g is J-antilinear; on the other hand, g admits a multiplicative factorization $g = uh$ in which $u \in U(V_J)$ is unitary and the J-linear part of h is self-adjoint. These two means of dissecting orthogonal transformations will be important to us when we consider implementability of the corresponding Bogoliubov automorphisms in the Fock representation.

We start from the rather general observation that any real-linear endomorphism g of V may be uniquely decomposed as the sum

$$g = C_g + A_g$$

of an endomorphism C_g that commutes with J (said to be *J-linear* or *complex linear*) and an endomorphism A_g that anticommutes with J (said to be *J-antilinear*, or simply *antilinear*). Quite explicitly, the complex-linear part C_g of g is given by

$$C_g = \tfrac{1}{2}(g - JgJ)$$

and the antilinear part A_g of g is given by

$$A_g = \tfrac{1}{2}(g + JgJ).$$

Now let g and h both be real-linear endomorphisms of V. From

$$C_{gh} + A_{gh} = gh = (C_g + A_g)(C_h + A_h)$$
$$= (C_g C_h + A_g A_h) + (C_g A_h + A_g C_h)$$

it follows upon taking linear and antilinear parts that
$$C_{gh} = C_g C_h + A_g A_h$$
and
$$A_{gh} = C_g A_h + A_g C_h.$$
As a special case, since $C_I = I$ and $A_I = 0$ we have the following result.

Theorem 3.1.1 *If g is a real-linear automorphism of V then*
$$I = C_{g^{-1}} C_g + A_{g^{-1}} A_g$$
$$0 = C_{g^{-1}} A_g + A_{g^{-1}} C_g.$$
$\square$

So far, we have not involved the inner product on V in our discussion. Doing so, it turns out that the linear and antilinear parts of bounded real-linear operators on V are well-behaved relative to the operator adjoint determined by the real inner product $(\cdot \mid \cdot)$. Indeed, denoting this adjoint operation by a star as usual, $J^* = -J$ so that if g is a bounded real-linear operator on V then $C_g^* = C_{g^*}$ and $A_g^* = A_{g^*}$. In particular, since $g^* = g^{-1}$ if $g \in O(V)$ is an orthogonal transformation, we deduce the following result.

Theorem 3.1.2 *If $g \in O(V)$ is an orthogonal transformation of V then $C_g^* = C_{g^{-1}}$ and $A_g^* = A_{g^{-1}}$ where adjunction is relative to $(\cdot \mid \cdot)$.*
$\square$

Here, the fact that adjunction takes place relative to the real inner product $(\cdot \mid \cdot)$ is important: adjunction relative to the complex inner product $\langle \cdot \mid \cdot \rangle = \langle \cdot \mid \cdot \rangle_J$ is a little different. If $g \in O(V)$ then the equation $C_g^* = C_{g^{-1}}$ remains valid without change but the relation between antilinear parts assumes the form that if $x, y \in V$ then $\langle A_g x \mid y \rangle = \langle A_{g^{-1}} y \mid x \rangle$ as is readily verified from Theorem 3.1.2 and the definition of $\langle \cdot \mid \cdot \rangle$. We record this remark as follows.

Theorem 3.1.3 *If $g \in O(V)$ is an orthogonal transformation of V and if $x, y \in V$ then*
$$\langle C_g x \mid y \rangle = \langle x \mid C_{g^{-1}} y \rangle$$
$$\langle A_g x \mid y \rangle = \langle A_{g^{-1}} y \mid x \rangle$$
$\square$

We can deduce one more set of conditions on the linear and antilinear parts of an orthogonal transformation g of V as follows. If we suppose that $v \in V$ then in light of Theorem 3.1.1 and Theorem 3.1.3 we see that

$$\langle C_g v \mid A_g v \rangle = \langle v \mid C_{g^{-1}} A_g v \rangle$$
$$= -\langle v \mid A_{g^{-1}} C_g v \rangle$$
$$= -\langle C_g v \mid A_g v \rangle$$

whence in fact $C_g v$ and $A_g v$ are orthogonal relative to $\langle \cdot \mid \cdot \rangle$ so that then

$$\|v\|^2 = \|gv\|^2$$
$$= \|C_g v + A_g v\|^2$$
$$= \|C_g v\|^2 + \|A_g v\|^2$$

on account of orthogonality.

Theorem 3.1.4 *If $g \in O(V)$ and if $v \in V$ then*
$$\langle C_g v \mid A_g v \rangle = 0$$
$$\|C_g v\|^2 + \|A_g v\|^2 = \|v\|^2.$$

$\square$

Incidentally, it is perhaps worth mentioning that we now have at hand two alternative criteria for membership in $O(V)$ of the bounded real-linear automorphism g of V. The first criterion is the converse to Theorem 3.1.4: if $\langle C_g v \mid A_g v \rangle = 0$ and $\|C_g v\|^2 + \|A_g v\|^2 = \|v\|^2$ for all $v \in V$ then $g \in O(V)$; indeed, satisfaction of these conditions implies that if $v \in V$ then

$$\|gv\|^2 = \|C_g v + A_g v\|^2 = \|C_g v\|^2 + \|A_g v\|^2 = \|v\|^2.$$

The second criterion relates to Theorem 3.1.1 and Theorem 3.1.2: if $C_g^* A_g + A_g^* C_g = 0$ and $C_g^* C_g + A_g^* A_g = I$ then $g \in O(V)$; indeed, these conditions respectively imply that if $v \in V$ then $\langle C_g v \mid A_g v \rangle = 0$ and $\|C_g v\|^2 + \|A_g v\|^2 = \|v\|^2$ as is readily checked.

We should also characterize membership in the unitary group $U(V_J)$ at some point; here is as appropriate a place as any.

Theorem 3.1.5 *The orthogonal transformation g of V lies in the unitary group $U(V_J)$ iff $Jg = gJ$ iff $C_g = g$ iff $A_g = 0$.*

Proof Equivalence of the three conditions $Jg = gJ$, $C_g = g$, $A_g = 0$ is valid for any real-linear operator g on V directly from the definitions. If $g \in O(V)$ and if $x, y \in V$ then

$$\langle gx \mid gy \rangle = (gx \mid gy) + \mathrm{i}(gx \mid Jgy)$$
$$= (x \mid y) + \mathrm{i}(x \mid g^{-1} Jgy)$$

so that g preserves $\langle \cdot \mid \cdot \rangle$ iff $g^{-1}Jg = J$ iff $Jg = gJ$ and we are done.

$\square$

Now, given any orthogonal transformation $g \in O(V)$ let us agree to write $N_g \subset V$ for the kernel of C_g and $R_g \subset V$ for the range of C_g. Since C_g is J-linear, both N_g and R_g are complex subspaces of $V = V_J$. Moreover, N_g is closed and $V = N_g \oplus N_g^\perp$ is an orthogonal decomposition relative to either $(\cdot \mid \cdot)$ or $\langle \cdot \mid \cdot \rangle$; indeed, the orthogonal complement of a complex subspace of V is the same relative to either $(\cdot \mid \cdot)$ or $\langle \cdot \mid \cdot \rangle$. Although R_g need not be closed, the identity $C_g^* = C_{g^{-1}}$ of Theorem 3.1.2 does at least ensure that

$$R_g^\perp = (\operatorname{ran} C_g)^\perp = \ker C_g^* = \ker C_{g^{-1}} = N_{g^{-1}}$$

and so

$$\overline{R}_g = R_g^{\perp\perp} = N_{g^{-1}}^\perp.$$

Of course, the restrictions of A_g and g itself to $N_g = \ker C_g$ coincide. In fact, if $v \in N_g$ then from Theorem 3.1.1 it follows that

$$C_{g^{-1}}(A_g v) = -A_{g^{-1}}(C_g v) = 0$$

so that

$$A_g(N_g) \subset N_{g^{-1}}$$

and similarly

$$A_{g^{-1}}(N_{g^{-1}}) \subset N_g.$$

Moreover, if again $v \in N_g$ then Theorem 3.1.1 implies that

$$v = (C_{g^{-1}}C_g + A_{g^{-1}}A_g)v = A_{g^{-1}}A_g v$$

whence

$$A_{g^{-1}}A_g \mid N_g = I$$

and likewise

$$A_g A_{g^{-1}} \mid N_{g^{-1}} = I.$$

Since Theorem 3.1.2 informs us that the antilinear maps A_g and $A_{g^{-1}}$ are mutual adjoints, we have established the following result.

Theorem 3.1.6 *If $g \in O(V)$ then its antilinear part A_g restricts to an antiunitary isomorphism*

$$A_g : N_g \to N_{g^{-1}}$$

whose inverse is $A_{g^{-1}}$.

$\square$

Of course, since also $C_g(N_g) = 0 \subset N_{g^{-1}}$ it follows that in fact

$g(N_g) \subset N_{g-1}$ and therefore $g(N_g^\perp) \subset N_{g-1}^\perp$ by orthogonality of g. Furthermore, $C_g(N_g^\perp) \subset R_g \subset N_{g-1}^\perp$ whence also $A_g(N_g^\perp) \subset N_{g-1}^\perp$. In summary, each of g, C_g and A_g has block form

$$\begin{bmatrix} * & 0 \\ 0 & * \end{bmatrix}$$

when we orthogonally decompose V as $N_g \oplus N_g^\perp$ initially and as $N_{g-1} \oplus N_{g-1}^\perp$ finally.

Theorem 3.1.7 *If $g \in O(V)$ then each of g, C_g and A_g maps N_g to N_{g-1} and maps $N_g^\perp$ to $N_{g-1}^\perp$.* $\qquad\square$

When the orthogonal transformation g has the special property that $C_g = C_g^*$ is self-adjoint, a number of simplifications occur. Indeed, from Theorem 3.1.2 it follows that $N_g = N_{g-1}$ so that each of g, C_g and A_g leaves the orthogonal decomposition $V = N_g \oplus N_g^\perp$ invariant.

Theorem 3.1.8 *If $g \in O(V)$ is such that C_g is self-adjoint, then each of g, C_g and A_g leaves invariant both N_g and $N_g^\perp$.* $\qquad\square$

Note that since $\overline{R}_g = N_{g-1}^\perp$ when $g \in O(V)$ is arbitrary, it follows here that the range of $C_g \mid N_g^\perp$ is dense in $N_g^\perp$ because $C_g \mid N_g$ is zero; this also follows from the fact that $C_g \mid N_g^\perp$ is injective and self-adjoint. Note also from Theorem 3.1.6 that A_g here restricts to an antiunitary automorphism of N_g.

We now direct our attention towards a multiplicative factorization of the orthogonal transformation $g \in O(V)$: in fact, we shall produce a unitary transformation $u \in U(V_J)$ with the property that the orthogonal transformation $u^{-1}g$ has self-adjoint J-linear part.

We begin with $g \in O(V)$ and take the polar decomposition of the bounded J-linear operator C_g on V: thus,

$$C_g = f_g \, |C_g|$$

where $|C_g|$ is the positive square root of $C_g^* C_g$ and where f_g is the partial isometry on V having initial space the closure $N_g^\perp$ of $\mathrm{ran} \, |C_g|$ and final space the closure $N_{g-1}^\perp$ of R_g. Recall from Theorem 3.1.6 that A_g restricts to an antiunitary isomorphism from N_g to N_{g-1}. In consequence, there exists a partial isometry e on V having N_g as its initial space and N_{g-1} as its final space: indeed, we may let e be zero on $N_g^\perp$ and be A_g precomposed with any antiunitary transformation on N_g.

These preparations in hand, we define a unitary transformation $u \in U(V_J)$ on V by putting

$$u = e + f_g.$$

Notice that $u \, |C_g| = C_g$ since if $v \in V$ then $|C_g| \, v \in N_g^\perp$ and therefore
$$u \, |C_g| \, v = f_g \, |C_g| \, v = C_g v.$$
We now have the splitting
$$g = C_g + A_g = u \, |C_g| + A_g$$
$$= u(\, |C_g| + u^* A_g)$$
in which u is unitary and in which the orthogonal transformation $|C_g| + u^* A_g$ has self-adjoint J-linear part.

Theorem 3.1.9 *Each orthogonal transformation $g \in O(V)$ admits a factorization*
$$g = u(\, |C_g| + u^* A_g)$$
in which $u \in U(V_J)$ is unitary and in which the J-linear part $|C_g|$ of $u^{-1}g \in O(V)$ is self-adjoint. $\qquad\Box$

This is our promised multiplicative factorization of the orthogonal transformation g. Unlike the additive decomposition $g = C_g + A_g$ it is not quite canonical. Although the partial isometry f_g is uniquely defined by polar decomposition, the partial isometry e may be precomposed with any unitary transformation on N_g. It is for this reason that we chose not to bestow a label g upon the unitary operator u. Needless to say, this lack of uniqueness is irrelevant for the use to which we shall put our multiplicative factorization in Section 3.3, where it will provide the final link in our complete determination of the Fock implementable Bogoliubov automorphisms of the C^* Clifford algebra.

We round off this section with a result that will also prove useful in our determination of Fock implementable Bogoliubov automorphisms.

Theorem 3.1.10 *If $g \in O(V)$ is such that A_g is compact then the restriction $C_g : N_g^\perp \to N_{g^{-1}}^\perp$ is an isomorphism.*

Proof Theorem 3.1.7 tells us that C_g maps $N_g^\perp$ to $N_{g^{-1}}^\perp$. The restricted operator $C_g : N_g^\perp \to N_{g^{-1}}^\perp$ is plainly injective; further, its range R_g is dense in $N_{g^{-1}}^\perp$. By the open mapping theorem, we need only show that C_g is bounded below on $N_g^\perp$. Suppose not: let $(v_n : n \geq 0)$ be a sequence of unit vectors in $N_g^\perp$ such that $\|C_g v_n\| \to 0$. Since A_g is compact, by passing to a subsequence and renumbering if need be, we may assume that $(A_g v_n : n \geq 0)$ converges; its limit w lies in $N_{g^{-1}}^\perp$ by Theorem 3.1.7. Now, Theorem 3.1.1 yields $\|C_{g^{-1}} A_g v_n\| = \|A_{g^{-1}} C_g v_n\|$ and Theorem 3.1.4 implies that $\|A_{g^{-1}} C_g v_n\| \leq \|C_g v_n\|$; thus
$$C_{g^{-1}} w = \lim_n C_{g^{-1}} A_g v_n = 0$$

and so $w \in N_{g^{-1}}$. At this point, we have $w \in N_{g^{-1}}^{\perp} \cap N_{g^{-1}} = 0$. However, from Theorem 3.1.4 we also have

$$\|w\|^2 = \lim_n \|A_g v_n\|^2 = \lim_n (1 - \|C_g v_n\|^2) = 1.$$

This is ridiculous.   □

In particular, the operator C_g has closed range $R_g = N_{g^{-1}}^{\perp}$. Sill supposing A_g to be compact, the kernel N_g of C_g is finite-dimensional: indeed, Theorem 3.1.6 tells us that the restriction $A_g : N_g \to N_{g^{-1}}$ is an antiunitary isomorphism; this restriction being also compact, its domain is necessarily of finite dimension. Theorem 3.1.6 also tells us that $R_g = N_{g^{-1}}^{\perp}$ is of finite codimension, equal to the dimension of N_g. In short, $C_g (= g - A_g)$ is a *Fredholm* operator of index zero.

3.2 Implementation and equivalence

We now introduce the principal topic of this chapter by making quite explicit the intimate relationship between two problems concerning Fock representations. First, the *implementation problem*: let J be a fixed choice of unitary structure on the real Hilbert space V and determine necessary and sufficient conditions on the orthogonal transformation $g \in O(V)$ in order that the Bogoliubov automorphism θ_g of $C[V]$ be *unitarily implementable* in the Fock representation π_J in the sense that there exists a unitary operator $U \in \operatorname{Aut} \mathbb{H}_J$ such that

$$a \in C[V] \quad \Rightarrow \quad \pi_J(\theta_g a) = U \pi_J(a) U^*.$$

Second, the *equivalence problem*: determine necessary and sufficient conditions on a pair of unitary structures J and K for V in order that their associated Fock representations π_J and π_K be *unitarily equivalent* in the sense that there exists a unitary isomorphism $T : \mathbb{H}_J \to \mathbb{H}_K$ such that

$$a \in C[V] \quad \Rightarrow \quad \pi_K(a) = T \pi_J(a) T^*.$$

As we shall see, the relationship between these problems is so intimate that a solution to the one yields a solution to the other. Our strategy will be to solve the implementation problem in the next section and deduce a solution to the equivalence problem in the section after that.

Here, we begin by recalling from Theorem 2.1.3 that the orthogonal group $O(V)$ acts transitively by conjugation on the set $\mathbb{U}(V)$ of unitary structures for V. Thus, if $J \in \mathbb{U}(V)$ is a fixed choice of unitary structure then any $K \in \mathbb{U}(V)$ has the form $K = gJg^{-1}$ for some orthogonal transformation g of V; of course g is not uniquely determined, the measure of indeterminacy being precomposition by an element of the unitary group $U(V_J)$. We shall maintain this notation throughout most of the present section: indeed, until after Theorem 3.2.3 in which we explicitly relate the problems of implementation and equivalence. Since two unitary structures are under consideration, we shall take care to label V by whichever is relevant.

Now, $g : V_J \rightarrow V_K$ is in fact a unitary isomorphism: indeed, this is how g was originally constructed from J and K in Theorem 2.1.3. By functoriality, g induces a unitary isomorphism $\bigwedge_g^n : \bigwedge^n(V_J) \rightarrow \bigwedge^n(V_K)$ for each natural number n: this is the identity when $n = 0$ and is given by the effect

$$\bigwedge_g^n (v_1 \wedge \ldots \wedge v_n) = g v_1 \wedge \ldots \wedge g v_n$$

on decomposables when $v_1, \ldots, v_n \in V$. These unitary isomorphisms together define a unitary isomorphism $\bigwedge_g = \bigoplus_{n \geq 0} \bigwedge_g^n$ from $H_J = H_J(V)$ to $H_K = H_K(V)$ which in turn extends by continuity to yield a unitary isomorphism (denoted by the same symbol) from $\mathbb{H}_J$ to $\mathbb{H}_K$. With a view to determining how $\bigwedge_g$ interacts with the Fock representations π_J and π_K we examine its interaction with creators and annihilators.

Theorem 3.2.1 *Let $K = gJg^{-1}$ with $J \in \mathbb{U}(V)$ and $g \in O(V)$. If $v \in V$ then*

$$c_K(gv) = \bigwedge_g \circ c_J(v) \circ \bigwedge_g^*$$

and

$$a_K(gv) = \bigwedge_g \circ a_J(v) \circ \bigwedge_g^*.$$

Proof Note that we are being careful, labelling creators and annihilators with the relevant unitary structures. If $v \in V$ and $\zeta \in H_J$ then

$$\bigwedge_g \circ c_J(v)(\zeta) = \bigwedge_g (v \wedge \zeta)$$

$$= gv \wedge \bigwedge_g \zeta$$

$$= c_K(gv) \circ \bigwedge_g (\zeta)$$

whence the equality

$$\bigwedge_g \circ c_J(v) = c_K(gv) \circ \bigwedge_g$$

holds on $\mathbb{H}_J$ by continuity. The equality

$$\bigwedge_g \circ a_J(v) = a_K(gv) \circ \bigwedge_g$$

for $v \in V$ may also be established by a direct computation; alternatively, it follows from the mutually adjoint nature of creators and annihilators. $\square$

Upon adding the two equations in the statement of this theorem, there results the fact that if $v \in V$ then

$$\pi_K(gv) = \bigwedge\nolimits_g \circ \pi_J(v) \circ \bigwedge\nolimits_g^*.$$

It now follows (for example, by virtue of the uniqueness clause in Theorem 1.2.4) that

$$a \in C[V] \quad \Rightarrow \quad \pi_K(\theta_g a) = \bigwedge\nolimits_g \circ \pi_J(a) \circ \bigwedge\nolimits_g^*$$

so that the unitary isomorphism $\bigwedge_g$ establishes a unitary equivalence from the Fock representation π_J to the transformed Fock representation $\pi_K \circ \theta_g$.

Theorem 3.2.2   *Let* $K = gJg^{-1}$ *with* $J \in \mathbb{U}(V)$ *and* $g \in O(V)$. *If* $a \in C[V]$ *then*

$$\pi_K(\theta_g a) = \bigwedge\nolimits_g \circ \pi_J(a) \circ \bigwedge\nolimits_g^*.$$

□

Thus, if $a \in C[V]$ then the diagram

$$
\begin{array}{ccc}
\mathbb{H}_J & \xrightarrow{\bigwedge_g} & \mathbb{H}_K \\
{\scriptstyle \pi_J(a)} \downarrow & & \downarrow {\scriptstyle \pi_K(\theta_g a)} \\
\mathbb{H}_J & \xrightarrow{\bigwedge_g} & \mathbb{H}_K
\end{array}
$$

is commutative. In terms of Fock states, if $a \in C[V]$ then

$$
\begin{aligned}
\sigma_K(\theta_g a) &= \langle \pi_K(\theta_g a)\Omega_K \mid \Omega_K \rangle_K \\
&= \langle \bigwedge\nolimits_g \pi_J(a) \bigwedge\nolimits_g^* \Omega_K \mid \Omega_K \rangle_K \\
&= \langle \pi_J(a)\Omega_J \mid \Omega_J \rangle_J \\
&= \sigma_J(a)
\end{aligned}
$$

since the unitary isomorphism $\bigwedge_g$ intertwines π_J with $\pi_K \circ \theta_g$ and maps Ω_J to Ω_K. Thus

$$\sigma_K \circ \theta_g = \sigma_J$$

and the action of $O(V)$ on Fock states by Bogoliubov automorphisms of $C[V]$ is transitive.

We have now reached the point at which we can relate the two problems outlined in the opening paragraph of this section.

In the one direction, let $U \in \operatorname{Aut}\mathbb{H}_J$ be a unitary operator implementing the Bogoliubov automorphism θ_g in the Fock representation π_J:

thus,
$$a \in C[V] \quad \Rightarrow \quad \pi_J(\theta_g a) = U\pi_J(a)U^*.$$
Defining $T = \bigwedge_g \circ U^*$ yields a unitary isomorphism from $\mathbb{H}_J$ to $\mathbb{H}_K$ such that if $a \in C[V]$ then
$$T\,\pi_J(a)\,T^* = \bigwedge_g U^*\pi_J(a)U \bigwedge_g^*$$
$$= \bigwedge_g \pi_J(\theta_g^{-1}a) \bigwedge_g^*$$
$$= \pi_K(a).$$
In the other direction, let $T : \mathbb{H}_J \to \mathbb{H}_K$ be a unitary isomorphism intertwining the Fock representations π_J and π_K: thus,
$$a \in C[V] \quad \Rightarrow \quad \pi_K(a) = T\,\pi_J(a)\,T^*.$$
Defining $U = T^* \circ \bigwedge_g$ yields a unitary operator on $\mathbb{H}_J$ such that if $a \in C[V]$ then
$$U\pi_J(a)U^* = T^* \bigwedge_g \pi_J(a) \bigwedge_g^* T$$
$$= T^*\,\pi_K(\theta_g a)\,T$$
$$= \pi_J(\theta_g a).$$
The precise relationship between the problems of implementation and equivalence is thus elucidated in the following result.

Theorem 3.2.3 *Let $K = gJg^{-1}$ with $J \in \mathbb{U}(V)$ and $g \in O(V)$. The equation $T \circ U = \bigwedge_g$ sets up a bijective correspondence between unitary isomorphisms $T : \mathbb{H}_J \to \mathbb{H}_K$ intertwining π_J with π_K and unitary operators $U \in \mathrm{Aut}\,\mathbb{H}_J$ implementing θ_g in π_J.* $\square$

Recall that if unitary structures $J, K \in \mathbb{U}(V)$ are given then the orthogonal transformation $g \in O(V)$ such that $K = gJg^{-1}$ is only determined up to precomposition by an element of $U(V_J)$ in general. Now, it is of course the case that the difference $K - J$ has operator norm at most two. We claim that if $\|K - J\| < 2$ then in fact g may be chosen in a canonical manner as the partially isometric factor in the polar decomposition of $I - KJ$.

Note first of all that the assumption $\|K - J\| < 2$ implies that the operator $I - KJ$ is invertible: indeed,
$$I - KJ = 2I - (K - J)J.$$
Note next that $I - KJ$ is also normal: indeed, by direct calculation, each of the products $(I - KJ)^*(I - KJ)$ and $(I - KJ)(I - KJ)^*$ equals $2I - (JK + KJ)$. Now consider the polar decomposition
$$I - KJ = gM.$$

Invertibility of $I - KJ$ ensures that the partial isometry g is in fact an orthogonal transformation and that the modulus $M = |I - KJ|$ is invertible. Further, the normality of $I - KJ$ implies that M and g commute whilst the identity $M^2 = 2I - (JK + KJ)$ makes it plain that M commutes with J and K. Thus, from

$$gMJ = (I - KJ)J = J + K$$

and

$$KgM = K(I - KJ) = K + J$$

we deduce that $gJ = Kg$ and complete the justification of our claim.

Theorem 3.2.4 *If $J, K \in \mathbb{U}(V)$ are such that $\|K - J\| < 2$ then the polar decomposition*

$$I - KJ = g|I - KJ|$$

provides a canonical $g \in O(V)$ with the property that $K = gJg^{-1}$. $\qquad\square$

As mentioned previously, we shall offer complete solutions to the problems of implementation and equivalence in the coming sections. For now, we draw the present section to a close by considering two special cases. The first concerns unitary transformations, which are automatically implemented in the corresponding Fock representation; this special case will play a significant rôle in our discussion of Fock implementability in the general case. The second yields simple direct evidence for the existence of inequivalent Fock representations in infinite dimensions.

For the first special case, let $J \in \mathbb{U}(V)$ be a fixed choice of unitary structure. If $g \in U(V_J) \subset O(V)$ is a unitary transformation of V_J then $J = gJg^{-1}$ so that $\bigwedge_g$ is in fact a unitary operator on the Fock space $\mathbb{H}_J$ such that $\bigwedge_g \Omega_J = \Omega_J$ and

$$a \in C[V] \quad \Rightarrow \quad \pi_J(\theta_g a) = \bigwedge_g \circ \pi_J(a) \circ \bigwedge_g^*$$

according to Theorem 3.2.2. Thus, unitary transformations of V_J are automatically implemented in the Fock representation π_J.

Theorem 3.2.5 *If $J \in \mathbb{U}(V)$ then each unitary transformation $g \in U(V_J)$ is canonically implemented in the Fock representation π_J by the unitary operator $\bigwedge_g \in \mathrm{Aut}\,\mathbb{H}_J$.* $\qquad\square$

In terms of Fock states, if $g \in U(V_J)$ then $\sigma_J \circ \theta_g = \sigma_J$ and the mechanics of the Gelfand-Naimark-Segal construction provides an alternative route to the canonical unitary operator on $\mathbb{H}_J$ fixing Ω_J and implementing θ_g in π_J.

For the second special case, let $J \in \mathbb{U}(V)$ again be fixed and suppose V to be infinite-dimensional. We contend that the Fock representations π_J and π_{-J} are not unitarily equivalent. To confirm this, let $T : \mathbb{H}_J \to \mathbb{H}_{-J}$ be a bounded linear operator such that $T \circ \pi_J(a) = \pi_{-J}(a) \circ T$ whenever $a \in C[V]$. In particular, if $v \in V$ then from Theorem 2.4.1 we deduce that

$$T \circ a_J(v) = T \circ \tfrac{1}{2} \pi_J(v + \mathrm{i} Jv)$$
$$= \tfrac{1}{2} \pi_{-J}(v + \mathrm{i} Jv) \circ T$$
$$= c_{-J}(v) \circ T.$$

Now, Theorem 2.3.7 implies that

$$c_{-J}(v)(T\,\Omega_J) = T\,(a_J(v)\Omega_J) = 0$$

and then Theorem 2.3.8 forces upon us the conclusion that $T\,\Omega_J = 0$. Thus, the operator T fails to be invertible, in support of our contention. In fact, the operator T itself is zero since the representation π_J is irreducible.

Theorem 3.2.6 *If $J \in \mathbb{U}(V)$ and if V is infinite-dimensional then the Fock representations π_J and π_{-J} are inequivalent.* $\qquad\qquad\square$

This result may be reformulated in terms of the implementation problem: the reformulation states that if $J \in \mathbb{U}(V)$ and if V is infinite-dimensional, then antiunitary transformations of V_J are not implemented in π_J.

3.3 Implementation

J is a fixed choice of unitary structure on the real Hilbert space V giving rise to the Fock representation π_J of the C^* Clifford algebra $C[V]$ on the Fock space $\mathbb{H}_J$ as usual. Our concern in the present section is with the solution of what we have called the *implementation problem*: explicitly, we determine necessary and sufficient conditions on the orthogonal transformation g of V in order that the Bogoliubov automorphism θ_g of $C[V]$ be unitarily implemented in π_J in the sense that there is a unitary operator U on $\mathbb{H}_J$ with

$$\pi_J(\theta_g a) = U \pi_J(a) U^*$$

for all $a \in C[V]$. It transpires that if $g \in O(V)$ then θ_g is unitarily implemented in π_J if and only if the antilinear part $A_g = \tfrac{1}{2}(g + JgJ)$ of g is a Hilbert-Schmidt operator; note that this is equivalent to the commutator $[g, J] = gJ - Jg$ being Hilbert-Schmidt, since this commutator equals

$2A_g J$. Put somewhat crudely, the Bogoliubov automorphism of $C[V]$ induced by $g \in O(V)$ is unitarily implemented in π_J if and only if g is close to being J-linear, closeness being measured in a Hilbert-Schmidt sense.

Now, the point behind the implementation problem is to find a necessary and sufficient condition on $g \in O(V)$ in order that the representation $\pi_J \circ \theta_g$ of $C[V]$ on $\mathbb{H}_J$ be unitarily equivalent to the Fock representation π_J itself. Viewing the problem in this way brings to mind Theorem 2.4.7: a representation π of $C[V]$ on a complex Hilbert space $\mathbb{H}$ is unitarily equivalent to π_J if and only if $\mathbb{H}$ contains a cyclic vector Ω with the J-vacuum property

$$v \in V \quad \Rightarrow \quad \pi(v + \mathrm{i}Jv)\Omega = 0.$$

Noting that any nonzero vector in $\mathbb{H}_J$ is automatically cyclic for $\pi_J \circ \theta_g$ since π_J is irreducible, it follows that in order for $\pi_J \circ \theta_g$ to be unitarily equivalent to π_J it is necessary and sufficient that $\mathbb{H}_J$ should contain a nonzero vector Ω with $\pi_J \circ \theta_g(v + \mathrm{i}Jv)\Omega = 0$ whenever $v \in V$.

The foregoing remarks lead us to consider a little more closely the operator $\pi_J \circ \theta_g(v + \mathrm{i}Jv) = \pi_J(gv + \mathrm{i}gJv)$ for $v \in V$ and $g \in O(V)$. A direct computation yields

$$\begin{aligned}
\pi_J \circ \theta_g(v + \mathrm{i}Jv) &= \pi_J(gv) + \mathrm{i}\pi_J(gJv) \\
&= c(gv) + a(gv) + \mathrm{i}c(gJv) + \mathrm{i}a(gJv) \\
&= c(gv) + a(gv) + c(JgJv) - a(JgJv) \\
&= c(gv + JgJv) + a(gv - JgJv) \\
&= 2c(A_g v) + 2a(C_g v)
\end{aligned}$$

in view of the J-linearity of $c = c_J$ and the J-antilinearity of $a = a_J$. We record the result of this computation for later reference.

Theorem 3.3.1 *If $g \in O(V)$ and if $v \in V$ then*

$$\tfrac{1}{2}\pi_J \circ \theta_g(v + \mathrm{i}Jv) = c_J(A_g v) + a_J(C_g v).$$

$\square$

Before addressing the implementation problem in full generality, we take a very special case: for $g \in O(V)$ with C_g invertible, we show that θ_g is unitarily implemented in π_J if A_g is Hilbert-Schmidt. In this case, the operator

$$Z_g = -A_g C_g^{-1}$$

is both antilinear and Hilbert-Schmidt; moreover, if $x, y \in V$ then

$$
\begin{aligned}
\langle Z_g x \mid y \rangle &= -\langle A_g C_g^{-1} x \mid y \rangle \\
&= -\langle A_{g^{-1}} y \mid C_g^{-1} x \rangle \\
&= -\langle C_{g^{-1}}^{-1} A_{g^{-1}} y \mid x \rangle \\
&= \langle A_g C_g^{-1} y \mid x \rangle \\
&= -\langle Z_g y \mid x \rangle
\end{aligned}
$$

on account of Theorem 3.1.1 and Theorem 3.1.3. Thus, Z_g actually lies in the space $\mathbb{S}(V_J)$ and so by Theorem 2.2.2 corresponds with an element $\zeta_g \in \bigwedge^2[V_J]$ according to the rule

$$
x, y \in V \quad \Rightarrow \quad \langle \zeta_g \mid x \wedge y \rangle = \langle Z_g x \mid y \rangle.
$$

We now claim that the quadratic exponential (or Gaussian) $\exp(\zeta_g) \in \mathbb{H}_J$ supplied by Theorem 2.2.5 is a (nonzero, hence cyclic) J-vacuum vector for $\pi_J \circ \theta_g$. Indeed, Theorem 2.3.10 enables us to infer that if $v \in V$ then

$$
\begin{aligned}
[\, c(A_g v) + a(C_g v) \,] &\exp(\zeta_g) \\
&= (A_g v) \wedge \exp(\zeta_g) + (Z_g C_g v) \wedge \exp(\zeta_g)
\end{aligned}
$$

which vanishes by definition of Z_g; the claim follows from this upon application of Theorem 3.3.1.

Theorem 3.3.2 *Let $g \in O(V)$ be such that C_g is invertible. If A_g is Hilbert-Schmidt then θ_g is unitarily implemented in π_J.* $\qquad\square$

Our complete solution to the implementation problem for $g \in O(V)$ will proceed in two stages: in the first, we consider a g for which the J-linear part C_g is self-adjoint; in the second, we consider a g that is entirely general.

To begin, let $g \in O(V)$ be an orthogonal transformation for which the J-linear part $C_g = C_g^*$ is self-adjoint. Recall from Theorem 3.1.8 that each of g, C_g and A_g preserves the orthogonal decomposition

$$
V = N_g \oplus N_g^\perp
$$

of V into the kernel N_g of C_g and its orthocomplement $N_g^\perp = \overline{R}_g$. For convenience, we shall write X in place of N_g and Y in place of $N_g^\perp$. Moreover, we shall write h and k for the orthogonal transformations induced by g upon restriction to X and Y respectively: thus, $h = g \mid X$ and $k = g \mid Y$. Note that if X and Y are viewed as complex Hilbert spaces in their own right, then

$$
\begin{aligned}
C_h &= C_g \mid X, & A_h &= A_g \mid X, \\
C_k &= C_g \mid Y, & A_k &= A_g \mid Y.
\end{aligned}
$$

Note also (as mentioned following Theorem 3.1.8) that $C_h = 0$, that A_h is an antiunitary operator on X and that $C_k = C_k^*$ is injective with range dense in Y.

Theorem 3.3.3 *Let $g \in O(V)$ be such that C_g is self-adjoint. If A_g is Hilbert-Schmidt then θ_g is unitarily implemented in π_J.*

Proof Choose a nonzero vector $\xi \in \bigwedge^m X$ where m is the complex dimension of $N_g = X$; this dimension is finite according to the remarks following Theorem 3.1.10. The restriction A_k of A_g to $N_g^\perp = Y$ is of course Hilbert-Schmidt, while Theorem 3.1.10 itself tells us that $C_k : Y \to Y$ is invertible; the proof of Theorem 3.3.2 therefore provides us with a vector $\eta \in \bigwedge^2[Y]$ such that the Gaussian $\exp(\eta)$ satisfies

$$y \in Y \quad \Rightarrow \quad [\, c(A_k y) + a(C_k y)\,]\exp(\eta) = 0.$$

Now, let $\zeta = \xi \wedge \exp(\eta) \in \mathbb{H}_J$. If $x \in X$ then

$$[\, c(A_g x) + a(C_g x)\,]\,\zeta = A_g x \wedge \xi \wedge \exp(\eta) = 0$$

since ξ lies in the top exterior power of $\ker C_g = X$. If $y \in Y$ then the above condition on the Gaussian $\exp(\eta)$ implies that

$$\begin{aligned}
[\, c(A_g y) + a(C_g y)\,]\,\zeta &= (-1)^m \xi \wedge c(A_g y)\exp(\eta) + a(C_g y)\xi \wedge \exp(\eta) \\
&\quad + (-1)^m \xi \wedge a(C_g y)\exp(\eta) \\
&= 0
\end{aligned}$$

since $\xi \in \bigwedge^m X$ and $C_g y \in Y$. By linearity, it follows that ζ is annihilated by $c(A_g v) + a(C_g v)$ whenever $v \in V$. Theorem 3.3.1 now informs us that ζ is a (plainly nonzero) J-vacuum vector for $\pi_J \circ \theta_g$ and Theorem 2.4.7 permits us to deduce that $\pi_J \circ \theta_g$ is unitarily equivalent to π_J. This concludes the proof.   □

Conversely, if $g \in O(V)$ is such that C_g is self-adjoint then θ_g being unitarily implemented in π_J implies that A_g is Hilbert-Schmidt. We approach this converse via a special case, assuming not only that C_g is self-adjoint but also that C_g is injective and so has dense range.

Thus, let U be a unitary operator on $\mathbb{H}_J$ such that if $a \in C[V]$ then $\pi_J(\theta_g a) = U\pi_J(a)U^*$ and note that $\Omega := U \cdot \Omega_J$ is then a (cyclic, since unit) J-vacuum vector for $\pi_J \circ \theta_g$. Decompose Ω into homogeneous components:

$$\Omega = \bigoplus_{n \geq 0} \omega_n$$

with $\omega_n \in \bigwedge^n[V_J]$ for each $n \in \mathbb{N}$. If $v \in V$ then taking components in the J-vacuum condition yields first that $a(C_g v)\omega_1 = 0$ and then that

$$n > 0 \quad \Rightarrow \quad a(C_g v)\omega_{n+1} + c(A_g v)\omega_{n-1} = 0.$$

The first of these conditions asserts that $\langle \omega_1 \mid C_g v \rangle = 0$ for all $v \in V$ and so forces $\omega_1 = 0$ since $\overline{R}_g = V$. By induction, the second forces all odd components of Ω to vanish. Induction on the second condition also shows that if $\omega_0 = 0$ then all even components of Ω vanish; Ω being nonzero, it follows that ω_0 is nonzero. Now put $\zeta = \omega_2/\omega_0$. The $n = 1$ component of the J-vacuum condition reads

$$v \in V \quad \Rightarrow \quad a(C_g v)\zeta + A_g v = 0$$

so that if $\zeta \in \bigwedge^2[V_J]$ corresponds to the Hilbert-Schmidt antiskew operator $Z \in \mathbb{S}(V_J)$ as in Theorem 2.2.2 then by Theorem 2.3.10 we deduce that $A_g = -Z \circ C_g$. Of course, it now follows that A_g is Hilbert-Schmidt.

We may now lift the additional assumption and prove the converse of Theorem 3.3.3 itself.

Theorem 3.3.4 *Let $g \in O(V)$ be such that C_g is self-adjoint. If θ_g is unitarily implemented in π_J then A_g is Hilbert-Schmidt.*

Proof Again, let $U \in \operatorname{Aut} \mathbb{H}_J$ implement θ_g in π_J and put $U \cdot \Omega_J = \Omega$. The J-vacuum condition on Ω implies that if $x \in X = N_g$ then

$$0 = [\, c(A_g x) + a(C_g x) \,]\, \Omega = A_g x \wedge \Omega$$

and therefore $X \wedge \Omega = 0$ since A_g restricts to an antiunitary automorphism of X by Theorem 3.1.6. Since Ω is nonzero, it follows by Theorem 2.3.9 that X has finite complex dimension m and that $\Omega = \xi \wedge \eta$ for some $\xi \in \bigwedge^m X$ and some $\eta \in \mathbb{H}_J(Y)$. The J-vacuum condition on Ω also implies that if $y \in Y = N_g^\perp$ then

$$0 = [\, c(A_g y) + a(C_g y) \,](\xi \wedge \eta)$$
$$= (-1)^m \xi \wedge [\, c(A_g y) + a(C_g y) \,]\eta$$

since $\xi \in \bigwedge^m X$ and $C_g y \in Y$. Recalling that $k = g \mid Y$ we deduce that

$$y \in Y \quad \Rightarrow \quad [\, c(A_k y) + a(C_k y) \,]\eta = 0$$

and therefore that $\eta \in \mathbb{H}_J(Y)$ is a cyclic J-vacuum vector for the representation $\pi_J \circ \theta_k$ of $C[Y]$. Since $C_k = C_k^*$ is injective on Y, from the special case considered prior to the theorem it follows that $A_k = A_g \mid Y$ is Hilbert-Schmidt. Since X is finite-dimensional, $A_h = A_g \mid X$ is trivially Hilbert-Schmidt. We conclude that A_g itself is a Hilbert-Schmidt operator, as required. □

We have now reached the point from which we can proceed to solve the implementation problem in full generality. Thus, let $g \in O(V)$ and recall from Theorem 3.1.9 that there exists a unitary $u \in U(V_J)$ with the property that the J-linear part $C_{u^*g} = |C_g|$ of $u^*g \in O(V)$ is self-adjoint. Note that gu^* then has self-adjoint J-linear part $u|C_g|u^*$ and has J-antilinear part A_gu^*. Recall further from Theorem 3.2.5 that the Bogoliubov automorphism θ_u of $C[V]$ is automatically implemented in π_J by the unitary operator $\bigwedge_u$ on $\mathbb{H}_J$. As a result of this recollection, θ_g is implemented in π_J if and only if θ_{gu^*} is implemented in π_J. Since A_gu^* is Hilbert-Schmidt if and only if A_g itself is Hilbert-Schmidt, we deduce from Theorem 3.3.3 and Theorem 3.3.4 the following solution to the implementation problem.

Theorem 3.3.5 *Let $g \in O(V)$. The Bogoliubov automorphism θ_g of $C[V]$ is unitarily implemented in the Fock representation π_J if and only if A_g is Hilbert-Schmidt.* □

As we remarked at the opening of this section, A_g is Hilbert-Schmidt if and only if the commutator $[g, J] = gJ - Jg$ is Hilbert-Schmidt.

After the preceding proof, a remark is certainly in order. Notice that for $g \in O(V)$ we chose to write $g = gu^*u$ with u unitary and the J-linear part of gu^* self-adjoint, rather than to write $g = uu^*g$. The reason behind this choice is that it enables a somewhat neater description of the transformed Fock vacuum $U \cdot \Omega_J$ when A_g is Hilbert-Schmidt and U is a unitary operator on $\mathbb{H}_J$ implementing θ_g in π_J. Indeed, the unitary operator $\bigwedge_u$ on $\mathbb{H}_J$ fixes the Fock vacuum Ω_J so that

$$U \cdot \Omega_J = U \circ \bigwedge_{u^*}(\Omega_J)$$

where $U \circ \bigwedge_{u^*}$ is now a unitary operator on $\mathbb{H}_J$ implementing θ_{gu^*} in π_J. Since gu^* has self-adjoint J-linear part, our rather explicit accounts in Theorem 3.3.2 and Theorem 3.3.3 may now be brought into play, with the following effect. Note that $C_{gu^*} = C_gu^*$ has kernel $u \cdot N_g$ with orthocomplement $u \cdot N_g^\perp$. If k denotes the restriction of gu^* to $u \cdot N_g^\perp$ then we define $\eta \in \bigwedge^2[u \cdot N_g^\perp]$ by

$$y_1, y_2 \in u \cdot N_g^\perp \quad \Rightarrow \quad \langle \eta \mid y_1 \wedge y_2 \rangle = -\langle A_k C_k^{-1}y_1 \mid y_2 \rangle$$

whereupon $U \cdot \Omega_J = \xi \wedge \exp(\eta)$ for a suitable (normalizing) choice of $\xi \in \bigwedge^m(u \cdot N_g)$ with m the complex dimension of N_g. Rather than record the detailed outcome of our deliberations we offer the following summary, sufficient for later reference.

Theorem 3.3.6 *Let $g \in O(V)$ be such that A_g is Hilbert-Schmidt and choose $u \in U(V_J)$ so that C_{gu^*} is self-adjoint. If U is a unitary operator* *on* $\mathbb{H}_J$ *implementing* θ_g *in* π_J *then*

$$U \cdot \Omega_J = \xi \wedge \exp(\eta)$$

for some ξ *in the top exterior power of* $u \cdot N_g$ *and some* $\eta \in \bigwedge^2[u \cdot N_g^{\perp}]$.

$\square$

This formula for the transformed Fock vacuum will prove useful to us in the final section of the present chapter, when we decide issues of parity.

Our solution to the implementation problem prompts us to introduce $O_J(V)$ to denote the collection of all orthogonal transformations g of V for which the commutator $[g, J]$ is a Hilbert-Schmidt operator: thus

$$O_J(V) = \{g \in O(V) : \|A_g\|_{\mathrm{HS}} < \infty\}.$$

Indeed, the content of Theorem 3.3.5 is that $O_J(V)$ comprises precisely all orthogonal transformations g of V for which the Bogoliubov automorphism θ_g of $C[V]$ is unitarily implemented in the Fock representation π_J. Of course, it follows from this that $O_J(V)$ is actually a subgroup of $O(V)$; however, a direct proof of this fact lies much nearer the surface.

Theorem 3.3.7 *$O_J(V)$ is a subgroup of $O(V)$ containing $U(V_J)$.*

Proof The inclusion $U(V_J) \subset O_J(V)$ is plain from Theorem 3.1.5: a unitary transformation of V_J has vanishing J-antilinear part. From Theorem 3.1.2 and the proof of Theorem 3.1.1 we deduce that if g and h lie in the orthogonal group on V then

$$A_{gh^{-1}} = A_g C_h^* + C_g A_h^*$$

whence it follows that if A_g and A_h are Hilbert-Schmidt then so is $A_{gh^{-1}}$.

$\square$

It is customary to refer to $O_J(V)$ as the *restricted orthogonal group* of the real Hilbert space V determined by the unitary structure J. This restricted orthogonal group is quite often denoted by $O_{\mathrm{res}}(V)$ when the choice of unitary structure is understood.

It is instructive to reformulate the restricted orthogonal group in terms of our alternative descriptions of a unitary structure as presented in Section 2.1. Before doing so, we note that if $g \in O(V)$ is extended to the complex Hilbert space $V^{\mathbb{C}}$ by complex linearity, then its extension g lies in the unitary group $U(V^{\mathbb{C}})$ and commutes with the canonical conjugation Σ of $V^{\mathbb{C}}$ over V. In fact, this process is reversible: it is readily checked that all elements of $U(V^{\mathbb{C}})$ commuting with Σ arise from $O(V)$ by complexification.

As in Theorem 2.1.4 we shall let F_J^+ and F_J^- denote the eigenspaces of J in $V^{\mathbb{C}}$ with eigenvalues $+i$ and $-i$ respectively. Let us agree to identify $F_J^{\pm}$ with V by means of the canonical isomorphisms

$$T^{\pm} : V \to F_J^{\pm} : v \mapsto (v \mp iJv)$$

so that T^+ is linear and T^- antilinear when V is made complex via J as usual. The isomorphisms T^+ and T^- allow us to regard maps among F_J^+ and F_J^- as maps from V to itself; we shall avail ourselves of this simplification when expressing operators in block form with respect to the orthogonal decomposition

$$V^{\mathbb{C}} = F_J^+ \oplus F_J^-.$$

Now, let g be an orthogonal (or any real-linear) transformation of V. If $v \in V$ then

$$g(v - iJv) = C_g v + A_g v - iC_g Jv - iA_g Jv$$
$$= (I - iJ)C_g v + (I + iJ)A_g v$$

since C_g is J-linear and A_g is J-antilinear. Thus

$$g(T^+ v) = T^+(C_g v) + T^-(A_g v)$$

and similarly

$$g(T^- v) = T^+(A_g v) + T^-(C_g v).$$

This means that (the complexification of) g has simplified block form

$$\begin{bmatrix} C_g & A_g \\ A_g & C_g \end{bmatrix}.$$

As in Theorem 2.1.6, we shall let P_J stand for orthogonal projection from $V^{\mathbb{C}}$ onto F_J^+; thus

$$P_J = \begin{bmatrix} I & 0 \\ 0 & 0 \end{bmatrix}$$

in (simplified) block form. Computing in terms of simplified block forms, it follows that

$$(I - P_J)gP_J = \begin{bmatrix} 0 & 0 \\ 0 & I \end{bmatrix} \begin{bmatrix} C_g & A_g \\ A_g & C_g \end{bmatrix} \begin{bmatrix} I & 0 \\ 0 & 0 \end{bmatrix} = \begin{bmatrix} 0 & 0 \\ A_g & 0 \end{bmatrix}$$

and likewise

$$P_J g(I - P_J) = \begin{bmatrix} 0 & A_g \\ 0 & 0 \end{bmatrix}.$$

The following alternative characterizations of restricted orthogonal transformations are now immediate.

Theorem 3.3.8 *Let $g \in O(V)$ be extended to $V^{\mathbb{C}}$ by complex-linearity. The following conditions are equivalent.*

(i) $g \in O_J(V)$.

(ii) g *has Hilbert-Schmidt off-diagonals in block form relative to* $V^{\mathbb{C}} =$
    $F_J^+ \oplus F_J^-$.

(iii) $(I - P_J)gP_J$ *or/and* $P_Jg(I - P_J)$ *is/are Hilbert-Schmidt.*

$\square$

3.4 Equivalence

Here, we solve the *equivalence problem*: to determine necessary and
sufficient conditions in order that the Fock representations of the C^* Clifford algebra $C[V]$ induced by a pair of unitary structures on the real
Hilbert space V should be unitarily equivalent. In addition, we present
some reformulations and consequences of our solution.

To begin, let us suppose that J and K are unitary structures on V.
In accordance with Theorem 2.1.3 let us choose an orthogonal transformation g of V such that $K = gJg^{-1}$. Now, Theorem 3.2.3 informs
us that π_J is unitarily equivalent to π_K if and only if θ_g is unitarily
implemented in π_J. Also, our solution to the implementation problem
in Theorem 3.3.5 implies that θ_g is unitarily implemented in π_J if and
only if the commutator $[g, J] = gJ - Jg$ is Hilbert-Schmidt. Lastly, the
identity $[g, J] = (K - J)g$ shows that $[g, J]$ is Hilbert-Schmidt if and only
if $K - J$ is Hilbert-Schmidt. In short, we have arrived at the following
solution to the equivalence problem.

Theorem 3.4.1 *Let J and K be unitary structures on V. The Fock
representations π_J and π_K of $C[V]$ are unitarily equivalent if and only
if the difference $K - J$ is Hilbert-Schmidt.* $\square$

Of course, an independent solution to the equivalence problem would
allow us to solve the implementation problem anew, by rearranging the
argument given above. For our purposes, it was simply more convenient
to solve the implementation problem first and then deduce a solution to
the equivalence problem.

As with the implementation problem, it is instructive to reformulate
our solution in terms of the alternative versions of unitary structure. We
refer as usual to Theorem 2.1.4 and Theorem 2.1.6 for the notation in
the following result.

Theorem 3.4.2 *If J and K are unitary structures on V then the
following conditions are equivalent:*

(i) *the Fock representations π_J and π_K are unitarily equivalent;*

(ii)   *the difference $P_K - P_J$ is a Hilbert-Schmidt operator;*

(iii)   *the composite linear operator*

$$F_K \subset V^{\mathbb{C}} \to \overline{F}_J$$

is Hilbert-Schmidt.

Proof The equivalence of (i) and (ii) is made manifest by Theorem 3.4.1 and the identity $K - J = 2i(P_K - P_J)$. The implication (ii)$\Rightarrow$(iii) follows at once from the remark that inclusion $F_K \subset V^{\mathbb{C}}$ composed with orthogonal projection $V^{\mathbb{C}} \to \overline{F}_J$ produces the linear operator $(I - P_J) \mid F_K = (P_K - P_J) \mid F_K$. The reverse implication (iii)$\Rightarrow$(ii) is a little less immediate: it follows from the fact that

$$P_K - P_J = (I - P_J)P_K - P_J(I - P_K)$$

$$= (I - P_J)P_K - \Sigma(I - P_J)P_K\Sigma$$

since $(I - P_J)P_K$ is zero on $\overline{F}_K$ and restricts to F_K as the operator $F_K \to \overline{F}_J$. $\square$

We may also reformulate Theorem 3.4.1 with reference to Fock states. The relation of unitary equivalence partitions the set of all states of the C^* Clifford algebra $C[V]$. Our solution to the equivalence problem tells us that this partition on the set $\{\sigma_J : J \in \mathbb{U}(V)\}$ of Fock states over $C[V]$ corresponds to the partition of its parameter set $\mathbb{U}(V)$ engendered by the relation under which $J, K \in \mathbb{U}(V)$ are equivalent if and only if $K - J$ is Hilbert-Schmidt.

We now proceed to examine some consequences of our solution to the equivalence problem, starting with a simple consequence of interest.

Theorem 3.4.3 *If $K \in \mathbb{U}(V)$ is obtained from $J \in \mathbb{U}(V)$ by changing sign on a closed subspace $Y \subset V_J$ then π_J and π_K are unitarily equivalent if and only if Y is finite-dimensional.*

Proof By hypothesis, the difference $K - J$ is equal to zero on $Y^\perp$ and to $-2J$ on Y. Since $\frac{1}{2}(K-J)$ is therefore a partial isometry, it is Hilbert-Schmidt if and only if its initial space Y is finite-dimensional. The result follows directly from Theorem 3.4.1. $\square$

In connection with this result, two subspaces W_1 and W_2 of a vector space are said to be *commensurable* if and only if their intersection $W_1 \cap W_2$ is of finite codimension in both W_1 and W_2. One direction of Theorem 3.4.3 may now be expressed in these terms as follows: if J

and K are unitary structures on V such that $F_J \subset V^{\mathbb{C}}$ and $F_K \subset V^{\mathbb{C}}$ are commensurable then the Fock representations π_J and π_K are unitarily equivalent.

A variation on the theme of the preceding result provides us with a simple demonstration of the fact that if V is infinite-dimensional then the C^* Clifford algebra $C[V]$ carries infinitely many inequivalent Fock representations. We first of all fix a choice J of unitary structure on V and let the cardinal d stand for the dimension of V_J as a complex Hilbert space; being infinite, d is also the cardinality of a complete orthonormal system for the real Hilbert space V. Now, any (infinite) set of cardinality d may be expressed as the disjoint union of d countably infinite subsets. Applying this remark to a complete orthonormal system for V_J produces a complex Hilbert space decomposition

$$V_J = \bigoplus_{j \in \mathcal{J}} V_j$$

in which each summand is a separable complex Hilbert space of infinite dimension and in which the indexing set $\mathcal{J}$ has cardinality d. For each $j \in \mathcal{J}$ we shall let J_j^+ denote the restriction of J to V_j and let J_j^- denote its negative. For each function $\varepsilon : \mathcal{J} \to \{+, -\}$ we may now define a unitary structure J^ε on V according to the formula

$$J^\varepsilon = \bigoplus_{j \in \mathcal{J}} J_j^{\varepsilon(j)}.$$

For example, if $\varepsilon \equiv +$ then $J^\varepsilon = J$ and if $\varepsilon \equiv -$ then $J^\varepsilon = -J$. Of course, as ε varies, the unitary structures J^ε constitute a cardinality 2^d subset of $\mathbb{U}(V)$. Moreover, if ε and ε' are distinct maps from $\mathcal{J}$ to $\{+, -\}$ and if $j \in \mathcal{J}$ is such that $\varepsilon(j) \neq \varepsilon'(j)$ then

$$\tfrac{1}{2}(J^\varepsilon - J^{\varepsilon'}) \mid V_j = J_j^{\varepsilon(j)}$$

whence the infinite-dimensionality of V_j implies that $J^\varepsilon - J^{\varepsilon'}$ is not Hilbert-Schmidt. Theorem 3.4.1 now implies that the 2^d unitary structures on V constructed above induce unitarily inequivalent Fock representations of the C^* Clifford algebra $C[V]$.

Theorem 3.4.4 *If the real Hilbert space V has infinite dimension d then $C[V]$ carries at least 2^d unitarily inequivalent Fock representations.*

$\square$

Notice in particular that the C^* Clifford algebra $C[V]$ has an uncountable number of inequivalent Fock representations even when the real Hilbert space V is separably infinite-dimensional.

The construction of inequivalent Fock representations in the proof of the preceding theorem is discrete and admittedly somewhat contrived. Given the fact that if V is infinite-dimensional then $C[V]$ carries a continuum of inequivalent Fock representations, we should attempt to produce a continuous curve of unitary structures on V such that the induced Fock representations are inequivalent. A successful attempt follows.

First of all, fix a unitary structure J on V. Choose any conjugation operator on the complex Hilbert space V_J and let W be its fixed space, so that $V = W \oplus JW$ is an orthogonal decomposition. Next, let K be any unitary structure on the real Hilbert space W and extend it to V by J-antilinearity, obtaining a new unitary structure on V which we continue to denote by K. Thus, J and K are anticommuting unitary structures on V.

We now define a norm continuous one-parameter subgroup of the orthogonal group on V by

$$g : \mathbb{R} \to O(V) : t \mapsto g_t = e^{tK}.$$

Of course, the corresponding curve of unitary structures $\{J_t : t \in \mathbb{R}\} \subset \mathbb{U}(V)$ given by

$$t \in \mathbb{R} \quad \Rightarrow \quad J_t = g_t J g_t^{-1}$$

is then also norm continuous. Observe that if $t \in \mathbb{R}$ then

$$e^{tK} - e^{-tK} = 2(\sin t)K$$

and

$$J e^{-tK} J^{-1} = e^{-tJKJ^{-1}} = e^{tK}$$

so that

$$[g_t, J] = [e^{tK}, J] = 2(\sin t)KJ.$$

In particular, the commutator $[g_t, J]$ is Hilbert-Schmidt if and only if the real parameter t is an integer multiple of π.

Finally, since $g : \mathbb{R} \to O(V)$ is a group homomorphism, elementary algebraic manipulations reveal that if $s, t \in \mathbb{R}$ then

$$[g_{t-s}, J] = (J_t - J_s)g_{t-s}$$

whence the difference $J_t - J_s$ is Hilbert-Schmidt if and only if the difference $t-s$ is an integer multiple of π. Taking Theorem 3.4.1 into account, we have established the following.

Theorem 3.4.5 *Let J and K be anticommuting unitary structures on the infinite-dimensional real Hilbert space V. If the norm continuous curve $\{J_r : r \in \mathbb{R}\} \subset \mathbb{U}(V)$ is given by*

$$r \in \mathbb{R} \quad \Rightarrow \quad J_r = e^{rK}Je^{-rK}$$

then the Fock representations of $C[V]$ induced by J_s and J_t are unitarily equivalent precisely when $t - s$ is an integer multiple of π. $\square$

Geometrically speaking, the idea behind this theorem is straightforward: to obtain the curve $\{J_r : r \in \mathbb{R}\}$ from J by applying rotations; the trick is to choose the rotations appropriately.

As a particular consequence of this theorem, restriction to the open interval $(-\frac{\pi}{2}, \frac{\pi}{2})$ results in an injective norm continuous curve

$$(-\tfrac{\pi}{2}, \tfrac{\pi}{2}) \to \mathbb{U}(V) : r \mapsto e^{rK} J e^{-rK} = e^{2rK} J$$

whose image consists of unitary structures for which the induced Fock representations are mutually inequivalent.

A little thought shows that we can say somewhat more than this. Again let J and K be anticommuting unitary structures on V; note that their product JK is a unitary structure on V that anticommutes with each. Taking a quaternionic cue, we realize that J and K actually determine a whole sphere of unitary structures on V: indeed if $p, q, r \in \mathbb{R}$ and $p^2 + q^2 + r^2 = 1$ then $pJ + qK + rJK$ is a unitary structure on V. As a subset of $\mathbb{U}(V)$ in the operator norm, this collection $S(J, K)$ of unitary structures on V is easily isomorphic to the standard sphere. However, any pair of points in $S(J, K)$ is infinitely separated in the Hilbert-Schmidt sense: in fact, if $p, q, r \in \mathbb{R}$ then a direct calculation reveals that if $v \in V$ is a unit vector then

$$\|(pJ + qK + rJK)v\|^2 = p^2 + q^2 + r^2$$

from which it is clear (upon summation over a complete orthonormal system for V) that $pJ + qK + rJK$ is Hilbert-Schmidt precisely when $p = q = r = 0$. Of course, it follows that the Fock representations of $C[V]$ induced by elements of the sphere $S(J, K) \subset \mathbb{U}(V)$ are unitarily inequivalent.

We make two remarks about this construction. First, the unitary structures of Theorem 3.4.5 constitute an equatorial circle of the sphere $S(J, K)$: explicitly, as t runs over the real numbers, the unitary structure

$$e^{tK} J e^{-tK} = (\cos 2t) J - (\sin 2t) JK$$

runs around the equator of $S(J, K)$ corresponding to K as pole. Second, the set of Fock representations of $C[V]$ arising from $S(J, K)$ still has only the cardinality of the continuum; Theorem 3.4.4 is of course superior in this respect.

3.5 Parity considerations

Recall from Section 2.5 that when restricted to the even C^* Clifford algebra, a Fock representation decomposes as the sum of two irreducibles, its even and odd components. Our aim in this section is to relate these even and odd Fock representations to matters of implementation and equivalence. Regarding implementation, we determine consequences of the way in which unitary operators implementing Bogoliubov automorphisms in a fixed Fock representation π_J of $C[V]$ interact with the corresponding grading operator Γ_J. Regarding equivalence, given unitary structures J and K on V we decide when either of the representations $\pi_J^{\pm}$ is equivalent to either of the representations $\pi_K^{\pm}$. As before, we deal first with implementation and then with equivalence.

Thus, let J be a fixed unitary structure on the real Hilbert space V. Let $g \in O_J(V)$ be a restricted orthogonal transformation of V so that A_g is Hilbert-Schmidt. Let U be a unitary operator on $\mathbb{H}_J$ implementing the Bogoliubov automorphism θ_g of $C[V]$ so that

$$v \in V \quad \Rightarrow \quad \pi_J(gv) = U\pi_J(v)U^*.$$

Theorem 2.5.1 reminds us that the grading operator Γ_J on $\mathbb{H}_J$ implements the grading automorphism γ of $C[V]$ in π_J. Accordingly, if $v \in V$ then

$$\Gamma_J U \Gamma_J \pi_J(v) \Gamma_J U^* \Gamma_J = -\Gamma_J U \pi_J(v) U^* \Gamma_J$$
$$= -\Gamma_J \pi_J(gv) \Gamma_J$$
$$= \pi_J(gv)$$

whence $\Gamma_J U \Gamma_J$ also implements θ_g in π_J. Irreducibility of π_J now forces $\Gamma_J U \Gamma_J$ to equal μU for some unitary scalar $\mu \in \mathbb{T}$. The fact that $\Gamma_J^2 = I$ forces $\mu^2 = 1$ so that $\mu = \pm 1$.

In consequence of these deliberations, if $U \in \operatorname{Aut} \mathbb{H}_J$ implements θ_g in π_J then either $\Gamma_J U \Gamma_J = U$ or $\Gamma_J U \Gamma_J = -U$: in the former case, U commutes with Γ_J and is said to be *even*; in the latter case, U anticommutes with Γ_J and is said to be *odd*. Notice that U is even if and only if it sends $\mathbb{H}_J^{\pm}$ to $\mathbb{H}_J^{\pm}$ whilst U is odd if and only if it sends $\mathbb{H}_J^{\pm}$ to $\mathbb{H}_J^{\mp}$. Thus, in order to decide whether U is even or odd, it suffices to decide whether the transformed Fock vacuum $U \cdot \Omega_J$ lies in $\mathbb{H}_J^+$ or $\mathbb{H}_J^-$. The key to making this decision is provided by Theorem 3.3.6: if $u \in U(V_J)$ is such that gu^* has self-adjoint J-linear part, then

$$U \cdot \Omega_J = \xi \wedge \exp(\eta)$$

for some ξ in the top exterior power of $u \cdot N_g$ and some $\eta \in \bigwedge^2[u \cdot N_g^{\perp}]$. This formula makes it clear that $U \cdot \Omega_J$ lies in $\mathbb{H}_J^+$ or $\mathbb{H}_J^-$ according to whether the complex dimension of $N_g = \ker C_g$ is even or odd respectively, since the Gaussian $\exp(\eta)$ lies in the even Fock space.

To summarize the outcome of the foregoing discussion, if $g \in O_J(V)$

then the *parity* (even or odd) of a unitary operator implementing θ_g in the Fock representation π_J is the same as the *parity* (even or odd) of the complex dimension of $N_g = \ker C_g$. In particular, notice that being even or odd is actually a property of the restricted orthogonal transformation itself: this is as it should be, since implementers are unique modulo scalar multiples and hence share the same parity. In order to formalize our findings more succinctly, let us agree that if $g \in O_J(V)$ then ε_g is to signify either $+$ or $-$ according to whether the complex dimension of N_g is even or odd respectively.

Theorem 3.5.1 *If $g \in O_J(V)$ and if U is a unitary operator on $\mathbb{H}_J$ that implements the Bogoliubov automorphism θ_g of $C[V]$ in π_J then*

$$\Gamma_J U \Gamma_J = \varepsilon_g U.$$

$\square$

This theorem has as an easy corollary the less than obvious fact that the map

$$\varepsilon : O_J(V) \to \{+, -\}$$

is actually a homomorphism of groups. We write $O_J^+(V)$ for the kernel of ε and $O_J^-(V)$ for its complement: thus, if $g \in O_J(V)$ is a restricted orthogonal transformation then $g \in O_J^+(V)$ precisely when $N_g = \ker C_g$ is even-dimensional and $g \in O_J^-(V)$ precisely when N_g is odd-dimensional. Notice that a unitary transformation $u \in U(V_J)$ automatically lies in $O_J^+(V)$: this is quite clear directly because $C_u = u$ is certainly injective, but also follows from the fact that θ_u is implemented in π_J by $\bigwedge_u$ and this commutes with $\bigwedge_{-I} = \Gamma_J$; see Theorem 3.1.5 and Theorem 3.2.5.

Now let J and K be a pair of unitary structures on V. We propose to consider the problem of deciding exactly when either of the representations $\pi_J^{\pm}$ is unitarily equivalent to either of $\pi_K^{\pm}$.

Our first claim along these lines is that if either of $\pi_J^{\pm}$ is unitarily equivalent to either of $\pi_K^{\pm}$ then in fact π_J is unitarily equivalent to π_K so that $K - J$ is Hilbert-Schmidt according to Theorem 3.4.1. To establish this claim, let $j, k \in \{+, -\}$ and suppose $T : \mathbb{H}_J^j \to \mathbb{H}_K^k$ to be a unitary isomorphism intertwining π_J^j with π_K^k as representations of $C^+[V]$. Select a unit vector $\Omega^j \in \mathbb{H}_J^j$ and put $\Omega^k = T\Omega^j \in \mathbb{H}_K^k$. Let ϕ stand for the state of $C[V]$ associated to π_J by the cyclic unit vector Ω^j so that

$$a \in C[V] \quad \Rightarrow \quad \phi(a) = \langle \pi_J(a)\Omega^j \mid \Omega^j \rangle;$$

similarly, let ψ be the state of $C[V]$ associated to π_K by the cyclic unit vector Ω^k. If $a \in C^+[V]$ then

$$
\begin{aligned}
\phi(a) &= \langle \pi_J(a)\Omega^j \mid \Omega^j \rangle \\
&= \langle \pi_J^j(a)\Omega^j \mid \Omega^j \rangle \\
&= \langle T^* \pi_K^k(a)T\,\Omega^j \mid \Omega^j \rangle \\
&= \langle \pi_K^k(a)\Omega^k \mid \Omega^k \rangle \\
&= \langle \pi_K(a)\Omega^k \mid \Omega^k \rangle \\
&= \psi(a).
\end{aligned}
$$

If $b \in C^-[V]$ then $\pi_J(b)$ maps $\mathbb{H}_J^j$ to $(\mathbb{H}_J^j)^\perp$ and $\pi_K(b)$ maps $\mathbb{H}_K^k$ to $(\mathbb{H}_K^k)^\perp$ so that $\phi(b) = 0 = \psi(b)$. It follows that $\phi = \psi$ and therefore that the associated cyclic representations π_J and π_K are unitarily equivalent, as was claimed. Having made good this first claim, we may now restrict attention to situations in which $K - J$ is Hilbert-Schmidt so that π_J and π_K are unitarily equivalent according to Theorem 3.4.1.

To proceed further with our investigations, we make contact with our recently completed discussion of implementation. Thus, let $g \in O(V)$ be an orthogonal transformation of V such that $gJg^{-1} = K$; note that g actually lies in $O_J(V)$ by virtue of the identity $[g, J] = (K - J)g$. The equation $T \circ U = \bigwedge_g$ in Theorem 3.2.3 sets up a bijection between the unitary isomorphisms $T : \mathbb{H}_J \to \mathbb{H}_K$ intertwining π_J with π_K and the unitary operators $U \in \operatorname{Aut}\mathbb{H}_J$ implementing θ_g in π_J. The relation $\Gamma_K \circ \bigwedge_g = \bigwedge_g \circ \Gamma_J$ ensures that $\bigwedge_g$ preserves parity, whence the correspondents T and U have the same parity properties: that is, $\Gamma_K T = \pm T\,\Gamma_J$ if and only if $\Gamma_J U \Gamma_J = \pm U$; equivalently, $\Gamma_K T\,\Gamma_J = \varepsilon_g T$ in the notation of Theorem 3.5.1. Thus: if $\varepsilon_g = +1$ then $\pi_J^\pm$ is unitarily equivalent to $\pi_K^\pm$ whilst if $\varepsilon_g = -1$ then $\pi_J^\pm$ is unitarily equivalent to $\pi_K^\mp$. All that remains is to eliminate the auxiliary g and express the sign ε_g directly in terms of J and K.

To this end, we make the claim that if linear and antilinear parts are taken relative to J then

$$
\ker(J + K) = g \cdot N_g = N_{g^{-1}}.
$$

Indeed, if $v \in V$ then

$$
\begin{aligned}
2C_g v &= gv - JgJv \\
&= gv - JKgv \\
&= -J(J + K)gv
\end{aligned}
$$

so that

$$
C_g v = 0 \quad \Leftrightarrow \quad (J + K)gv = 0.
$$

This deals with the first equality; the second is implicit in Theorem 3.1.7.

Of course, the foregoing remarks apply to situations in which g is unrestricted. In the present situation, $g \in O_J(V)$ is restricted so that N_g is finite-dimensional and the foregoing analysis yields that the complex dimension of N_g is half the real dimension of $\ker(J + K)$. We conclude that $\varepsilon_g = +$ precisely when $\frac{1}{2} \dim \ker(J + K)$ is even and that $\varepsilon_g = -$ precisely when $\frac{1}{2} \dim \ker(J + K)$ is odd.

Taken together, the preceding three paragraphs decide the matter of unitary equivalence for the even and odd Fock representations arising from a pair of unitary structures.

Theorem 3.5.2 *Let $J, K \in \mathbb{U}(V)$. If either of $\pi_J^\pm$ is unitarily equivalent to either of $\pi_K^\pm$ then $K - J$ is Hilbert-Schmidt. If $K - J$ is Hilbert-Schmidt then $\pi_J^\pm$ is unitarily equivalent to $\pi_K^\pm$ or to $\pi_K^\mp$ according to whether $\frac{1}{2} \dim \ker(J + K)$ is even or odd respectively.* $\qquad\square$

We end the present section by developing this theorem in two directions: the first a reformulation, the second an illustrative special case.

First, we reformulate the more delicate part of Theorem 3.5.2 as follows.

Theorem 3.5.3 *Let $J, K \in \mathbb{U}(V)$. If the composite linear operator*

$$F_K \subset V^{\mathbb{C}} \to \overline{F}_J$$

is Hilbert-Schmidt, then $\pi_J^\pm$ is unitarily equivalent to $\pi_K^\pm$ or to $\pi_K^\mp$ according to whether the (complex) dimension of $F_K \cap \overline{F}_J$ is even or odd respectively.

Proof Theorem 3.5.2 reduces our task to that of relating $F_K \cap \overline{F}_J$ and the kernel of $J + K$. If $v \in V$ then

$$2\mathrm{i}P_J(v - \mathrm{i}Kv) = (I - \mathrm{i}J)(Jv + Kv)$$

so that the diagram

$$
\begin{array}{ccc}
V & \xrightarrow{\ J+K\ } & V \\
\downarrow & & \downarrow \\
F_K & \xrightarrow{\ P_J|F_K\ } & F_J
\end{array}
$$

commutes, where the vertical isomorphisms are the standard identifications as in Theorem 2.1.8 for example. The equation

$$F_K \cap \overline{F}_J = \ker(P_J \mid F_K)$$

now renders the theorem transparent. $\qquad\square$

Second, we consider a special case of Theorem 3.5.2 referring to the context of Theorem 3.4.3.

Theorem 3.5.4 *If $K \in \mathbb{U}(V)$ is obtained from $J \in \mathbb{U}(V)$ by changing sign on a finite-dimensional subspace $Y \subset V_J$ then $\pi_J^\pm$ is unitarily equivalent to $\pi_K^\pm$ or to $\pi_K^\mp$ according to whether the (complex) dimension of Y is even or odd respectively.*

Proof This follows directly from Theorem 3.5.2 upon observing that the kernel of $J + K$ is precisely Y.   □

Remarks

The restricted orthogonal group

The present chapter can justifiably be said to revolve around the restricted orthogonal group $O_J(V)$ of the real Hilbert space V determined by a choice J of unitary structure. Appreciably more is known about this group than we have been able to include. Thus, it is naturally provided with topologies relative to which it is a topological group. One of these is a metric topology, for the metric given by the formula

$$d(g, h) = \|C_g - C_h\|_{\mathrm{OP}} + \|A_g - A_h\|_{\mathrm{HS}}.$$

Another is defined in terms of net convergence by stipulating that $g_j \to g$ if and only if $C_{g_j} \to C_g$ in the strong operator topology and $A_{g_j} \to A_g$ in the Hilbert-Schmidt topology. For both of these, the fact that $O_J(V)$ is a topological group follows from the identities

$$C_{gh^{-1}} = C_g C_h^* + A_g A_h^*$$
$$A_{gh^{-1}} = A_g C_h^* + C_g A_h^*.$$

It turns out that in both topologies, $O_J(V)$ has two components: these are precisely $O_J^+(V)$ and $O_J^-(V)$. These components are separated (as in Theorem 3.5.1) by the homomorphism ε ascribing to $g \in O_J(V)$ the sign $(-1)^{n_g}$ where n_g is the complex dimension of $N_g = \ker C_g$; thus, ε is an index map for the restricted orthogonal group. Moreover, if V is infinite-dimensional then the identity component $O_J^+(V)$ of the restricted orthogonal group is simply-connected. For further details, see [17] [20] [21]; see also [3].

Unitary implementers

Let us agree to denote by $\mathrm{Imp}_J(V)$ the group of all unitary operators on Fock space implementing Bogoliubov automorphisms of $C[V]$ in the Fock representation π_J. Our solution to the implementation problem shows that there is a canonical surjective group homomorphism from $\operatorname{Imp}_J(V)$ to $O_J(V)$ whose kernel is the circle $\mathbb{T}$ of unitary scalars since π_J is irreducible: thus, there is a central short exact sequence

$$1 \to \mathbb{T} \to \operatorname{Imp}_J(V) \to O_J(V) \to 1$$

of groups. A wealth of information on this topic is contained in the extensive article [3] by Araki, who refers to $\operatorname{Imp}_J(V)$ as a current group. Araki shows that $\operatorname{Imp}_J(V)$ is itself a topological group in the strong operator topology and has exactly two components $\operatorname{Imp}_J^+(V)$ and $\operatorname{Imp}_J^-(V)$ sitting over $O_J^+(V)$ and $O_J^-(V)$ respectively. Araki also discusses a Lie algebra $o_J(V)$ of $O_J(V)$ and explicitly computes the cocycle of its infinitesimal projective representation in Fock space: this is seen to be both a Schwinger term and a cyclic cocycle in the sense of Connes. As a result of this analysis, it follows that the short exact sequence for $\operatorname{Imp}_J(V)$ does not split. Moreover, the central circle extension $\operatorname{Imp}_J(V)$ cannot be cut down to a (connected) double cover of $O_J(V)$ since $O_J^+(V)$ is simply-connected when V is infinite-dimensional.

Universal implementation

Another natural matter not touched upon in the text is the following question: which Bogoliubov automorphisms of the C^* Clifford algebra are implemented in every Fock representation? The answer to this question is as follows: if $g \in O(V)$ then the Bogoliubov automorphism θ_g of $C[V]$ is implemented in π_J for all $J \in \mathbb{U}(V)$ precisely when either $g - I$ or $g + I$ is Hilbert-Schmidt. The sufficiency of this condition is transparent, since if $J \in \mathbb{U}(V)$ then of course $[g, J] = [g \mp I, J]$. Necessity is less clear: our solution to the implementation problem shows that if θ_g is universally Fock implemented then $[g, J]$ must be Hilbert-Schmidt when J is any unitary structure; spectral theory completes the argument. Essentially this approach to the question is adopted by Shale & Stinespring [81] and by Araki [2]. An alternative argument establishing necessity (for separable V of infinite dimension) begins by noting that the set of $g \in O(V)$ for which θ_g is universally Fock implemented is a proper normal subgroup of the orthogonal group, whence such g differ from either $+I$ or $-I$ by a compact operator according to a theorem of de la Harpe [44]; this alternative argument then proceeds by a (compact operator) spectral decomposition such as that to be found in Section 4.2.

Pfaffian construction

The book [67] by Pressley & Segal offers a rather different approach to some of the material in this chapter. Here, the group $\mathrm{Imp}\,^+_J(V)$ associated to $J \in \mathbb{U}(V)$ is constructed without direct reference to the Fock representation π_J and in fact without explicit reference to Clifford algebras. One consequence of the abstract (yet decidedly elegant) definition of $\mathrm{Imp}\,^+_J(V)$ given in Chapter 12 of [67] is that its 'spin' representation must be fashioned separately: it does not come automatically. It would take us too far afield to describe the Pressley & Segal construction in any detail. We merely note that the construction makes use of (relative) Pfaffians for Hilbert space operators, details on which may be found in [48]. Briefly, if $X, Y \in \mathbb{S}(V_J)$ are Hilbert-Schmidt antiskew operators on V_J then $I - XY$ of course has a Fredholm determinant, of which the *Pfaffian* $\mathrm{Pf}\,(X, Y)$ is a canonical square root.

History and miscellany

The understanding that unitarily inequivalent Fock representations exist and are important in infinite dimensions arose in the 1950s from work of Segal, van Hove, Haag, Friedrichs and others on quantum field theory. A variety of representations of the canonical anticommutation relations was displayed in the paper [39] of Gårding & Wightman, in which was proposed a classification based on properties of a number operator. We shall make no attempt here to survey the relevant literature on quantum field theory; however, we note that the significance of the Hilbert-Schmidt condition for implementation and equivalence is already expressed quite clearly in the classic text [38] by Friedrichs. For further information on quantum field theoretic aspects, see for example the references [7] [35] [41] [79] [83].

In a different form, the solution to the implementation problem appears in the paper [81] by Shale & Stinespring. The equivalence problem was solved in [57] by Manuceau & Verbeure, working with Fock states rather than with Fock representations. The idea to handle implementation and equivalence together was taken up by Araki in [3]. Accounts formulated more directly in terms of Fock space (and hence more in line with our exposition) were given by Berezin [12] and subsequently by Ruijsenaars [75] [76]. For further accounts of implementation and equivalence, see for example the references [4] [29] [52] [58] [66] [69].

4

SPIN GROUPS

In this closing chapter we consider Bogoliubov automorphisms once again, this time with a view to determining precisely which are inner as automorphisms of the various Clifford algebras over the real inner product space V. We begin in §1 with an examination of the plain complex Clifford algebra $C(V)$. Here, it turns out that if $g \in O(V)$ then the Bogoliubov automorphism θ_g of $C(V)$ is inner precisely when either $g - I$ has finite rank and $\ker(g + I)$ has even dimension or $g + I$ has finite rank and $\ker(g - I)$ has odd dimension. These conditions describe the full orthogonal group $O(V)$ when V is even-dimensional and the special orthogonal group $SO(V)$ when V is odd-dimensional; they are of course rather restrictive when V is infinite-dimensional. In §2 we suppose V to be an infinite-dimensional real Hilbert space and determine that if $g \in O(V)$ then the Bogoliubov automorphism θ_g of the C^* Clifford algebra $C[V]$ is inner precisely when either $g - I$ is of trace class and $\ker(g + I)$ is even-dimensional or $g + I$ is of trace class and $\ker(g - I)$ is odd-dimensional. Here, justification proceeds by cases and makes use of the Fock representations developed earlier. In §3 we again suppose V to be an infinite-dimensional real Hilbert space and find that if $g \in O(V)$ then the Bogoliubov automorphism θ_g of the vN Clifford algebra $\mathcal{A}[V]$ is inner precisely when either $g - I$ is Hilbert-Schmidt and $\ker(g + I)$ even-dimensional or $g + I$ is Hilbert-Schmidt and $\ker(g - I)$ odd-dimensional. Our proof of this fact makes surprisingly effective use of the theory of Fock representations. In each of these cases, the appropriate Clifford algebra harbours a group of unitaries that constitutes a double cover of a corresponding subgroup of the orthogonal group: the relevant subgroup of $O(V)$ comprises all $g \in O(V)$ for which θ_g is inner as an automorphism of the appropriate Clifford algebra; the double cover is formed from certain unitary elements implementing the inner automorphism. By analogy with terminology originating in finite dimensions, we refer to these double covers of orthogonal subgroups as pin groups and spin groups. In the Remarks, we assemble notes on alternative proofs, on topological aspects of spin groups, on outer invariants of Bogoliubov automorphisms, on the twisted adjoint representation, and on points of history. We should perhaps point out that although spin groups lend their name to the title of this chapter, we do not dwell on them: our main concern is with a description of inner Bogoliubov automorphisms.

4.1 Spin groups

We open our account of spin groups by considering those that arise within the plain complex Clifford algebra; spin groups within the C^* Clifford algebra and the vN Clifford algebra will be dealt with in the following sections.

Thus, let V be a real inner product space and $C(V)$ its complex Clifford algebra. Our primary aim in this section is to determine the orthogonal transformations $g \in O(V)$ for which the Bogoliubov automorphism θ_g of $C(V)$ is inner. We shall see that the dimension of V gives rise to a trichotomy. If the dimension of V is even, then every Bogoliubov automorphism of $C(V)$ is inner; if the dimension of V is odd, then the Bogoliubov automorphism θ_g is inner precisely when g lies in the special orthogonal group. The situation for an infinite-dimensional V is more complicated: θ_g is inner if and only if either $g - I$ has finite rank and $\ker(g+I)$ has even dimension or $g+I$ has finite rank and $\ker(g-I)$ has odd dimension.

Let us begin by supposing that the dimension of V is finite and that $g \in O(V)$ is an arbitrary orthogonal transformation. According to the spectral theorem, the real inner product space V admits a g-stable orthogonal decomposition

$$V = W^+ \oplus W \oplus W^-$$

where

$$W^\pm = \ker(g \mp I)$$

and where W is an orthogonal sum of planes on each of which g acts by a definite rotation: say

$$W = W_1 \oplus \cdots \oplus W_n$$

where if $j \in \mathbf{n}$ then $g_j = g \mid W_j$ is rotation through an angle φ_j in the range $0 < |\varphi_j| < \pi$. In terms of matrices, V has an orthonormal basis relative to which the matrix of g has block form

$$\begin{bmatrix} I & & & & \\ & g_1 & & & \\ & & \ddots & & \\ & & & g_n & \\ & & & & -I \end{bmatrix}$$

where if $j \in \mathbf{n}$ then

$$g_j = \begin{bmatrix} \cos \varphi_j & -\sin \varphi_j \\ \sin \varphi_j & \cos \varphi_j \end{bmatrix}.$$

Incidentally, it is worth pointing out that

$$\det g = (-1)^{\dim W^-}$$

since each of $g_1, \ldots, g_n$ is a rotation. In particular, g lies in the *special orthogonal group*

$$SO(V) = O^+(V) = \{g \in O(V) : \det g = 1\}$$

if and only if $W^- = \ker(g+I)$ is even-dimensional; we write $O^-(V)$ for the complement of $O^+(V)$ in $O(V)$.

The form taken by the spectral decomposition makes it clear that we should pay special attention to the planes on which g acts by rotation. To simplify notation for now, let us suppose W to be a plane on which $g \in O(V)$ acts by rotation through the angle φ: thus

$$g \cdot x = \cos \varphi \cdot x + \sin \varphi \cdot y$$
$$g \cdot y = -\sin \varphi \cdot x + \cos \varphi \cdot y$$

for $\{x, y\}$ an orthonormal basis of W relative to which g has matrix

$$[g] = \begin{bmatrix} \cos \varphi & -\sin \varphi \\ \sin \varphi & \cos \varphi \end{bmatrix}.$$

Again for convenience, write $c = \cos(\frac{1}{2}\varphi)$ and $s = \sin(\frac{1}{2}\varphi)$. Consider the unitary element

$$u = c - sxy \in C^+(W).$$

Since x and y anticommute by virtue of the Clifford relations, it follows from the standard double angle formulae that

$$uxu^* = (c - sxy)(cx + sy)$$
$$= (c^2 - s^2)x + 2scy$$
$$= \cos \varphi \cdot x + \sin \varphi \cdot y$$
$$= g \cdot x$$

and

$$uyu^* = (cy - sx)(c + sxy)$$
$$= -2scx + (c^2 - s^2)y$$
$$= -\sin\varphi \cdot x + \cos\varphi \cdot y$$
$$= g \cdot y.$$

These computations show that θ_g is inner as an automorphism of $C(W)$ and indeed is implemented by the even unitary u.

We now revert to the original notation, in which W is the orthogonal sum of the planes $W_1, \ldots, W_n$. For each $j \in \mathbf{n}$ we apply to W_j the construction of the previous paragraph, thereby obtaining an even unitary $u_j \in C^+(W_j)$ with the property that $u_j w u_j^* = g_j w = gw$ whenever $w \in W_j$. Since the planes $W_1, \ldots, W_n$ are orthogonal, it follows from the Clifford relations that the even unitary elements $u_1, \ldots, u_n$ commute. Moreover, their product $u_1 \ldots u_n$ is an even unitary in $C^+(W)$ that implements θ_g as an automorphism of $C(W)$. In order to decide whether or not θ_g is inner as an automorphism of $C(V)$ itself, it remains for us to examine behaviour on the orthocomplement $W^\perp = W^+ \oplus W^-$. We do this as follows.

Suppose that $g \in SO(V) = O^+(V)$ is a special orthogonal transformation: thus, $W^- = \ker(g+I)$ is even-dimensional. Let $\omega^- \in C^+(W^-)$ be the product of the vectors in an orthonormal basis for W^-: this even unitary element implements the grading automorphism $\gamma = \theta_{-I}$ on $C(W^-)$ by Theorem 1.1.10 and commutes with each element of $C(W^+ \oplus W)$ by the Clifford relations. Carrying over notation from the previous paragraph, the product $u = u_1 \ldots u_n \, \omega^-$ is now an even unitary element of $C^+(W \oplus W^-)$ that implements θ_g on $C(W \oplus W^-)$ and commutes with each element of $C(W^+)$. Of course, this means that u actually implements θ_g as an automorphism of $C(V)$ itself.

Thus, if $g \in SO(V) = O^+(V)$ is a special orthogonal transformation then θ_g is an inner automorphism of $C(V)$ regardless of dimensional parity. As we shall see, if $g \in O^-(V)$ then θ_g is inner if V is even-dimensional and not otherwise.

Theorem 4.1.1 *If V is even-dimensional then θ_g is inner whenever $g \in O(V)$.*

Proof For $v \in V$ any unit vector, let h_v signify reflection in the hyperplane orthogonal to v; note that each of $\pm h_v$ has determinant -1 and so lies in $O^-(V)$. The Clifford relations imply that if $z \in v^\perp$ then $vzv^* = -z$ whilst of course $vvv^* = v$; thus v implements θ_{-h_v} as an automorphism of $C(V)$. Now, if $g \in O^-(V)$ then $-h_v g \in O^+(V)$ and so

$\theta_{-h_v g} = \theta_{-h_v} \theta_g$ is inner; since θ_{-h_v} is already inner, we conclude that θ_g is also inner. $\qquad \square$

Note that here, if $g \in SO(V) = O^+(V)$ then θ_g is implemented by an even unitary whilst if $g \in O^-(V)$ then θ_g is implemented by an odd unitary.

Theorem 4.1.2 *If V is odd-dimensional then θ_g is inner precisely when $g \in SO(V) = O^+(V)$.*

Proof We need only show that if $g \in O^-(V)$ then θ_g is not inner. Note first that since V and W^- are odd-dimensional, $W^+ = \ker(g - I)$ is even-dimensional. From $-g^{-1} + I = g^{-1}(g - I)$ it now follows that $\ker(-g^{-1}+I)$ is even-dimensional, whence $-g^{-1} \in O^+(V)$ and therefore $\theta_{-g^{-1}}$ is inner. Now, if θ_g itself were inner, then so would be the grading automorphism $\gamma = \theta_{-I} = \theta_{-g^{-1}}\theta_g$. This is contrary to fact: see the closing paragraph in Section 1.1. $\qquad \square$

In this case, a retrospective glance reveals that if $g \in SO(V)$ then our construction provided θ_g with an even unitary implementer u. However, the product ω of vectors in an orthonormal basis for V is an odd unitary in the centre of $C(V)$ and hence implements the identity automorphism. It follows that θ_g is also implemented by an odd unitary, namely ωu.

This concludes what we wish to say at present regarding inner Bogoliubov automorphisms of the complex Clifford algebra in finite dimensions; we turn next to a consideration of the infinite-dimensional situation.

Thus, let the real inner product space V now be infinite-dimensional. Quite generally, let θ be an automorphism of $C(V)$ that commutes with the grading automorphism γ. Suppose θ to be inner, implemented by the invertible element $u \in C(V)$ say. If $a \in C(V)$ then

$$\gamma(u)a\gamma(u)^{-1} = \gamma\big(u\gamma(a)u^{-1}\big)$$
$$= \gamma\big(\theta(\gamma(a))\big) = \theta\big(\gamma\gamma(a)\big)$$
$$= \theta(a) = uau^{-1}$$

whence $\gamma(u) = \mu u$ for some (nonzero) scalar $\mu \in \mathbb{C}$ since $C(V)$ has scalar centre by Theorem 1.1.14. Since $\gamma^2 = I$ it follows that $\mu^2 = 1$ and therefore that $\mu = \pm 1$. In case $\gamma(u) = u$ we shall refer to the automorphism θ as being *even* and in case $\gamma(u) = -u$ we shall refer to θ as being *odd*; note that this is sensible, since all invertibles implementing θ are proportional and hence have the same parity. In either case, Theorem 1.1.13 tells us that the implementer u may be found in $C(M)$

for some finite-dimensional subspace $M \in \mathcal{F}(V)$ of V. If $u \in C^+(M)$ is even and if $a \in C(M^\perp)$ then $ua = au$ so that $\theta(a) = uau^{-1} = a$; it follows that $\theta \mid C(M^\perp) = I$ in this case. If $u \in C^-(M)$ is odd and if $a \in C(M^\perp)$ then $ua = \gamma(a)u$ so that $\theta(a) = uau^{-1} = \gamma(a)$; thus $\theta \mid C(M^\perp) = \gamma$ in this case. We summarize the salient points from this discussion as follows.

Theorem 4.1.3 *Let V be infinite-dimensional. Let θ be an inner automorphism of $C(V)$ commuting with γ and having implementer in $C(M)$ for some $M \in \mathcal{F}(V)$. If θ is even then $\theta \mid C(M^\perp) = I$ whilst if θ is odd then $\theta \mid C(M^\perp) = \gamma$.*   □

All of this applies in particular to Bogoliubov automorphisms, since they certainly commute with the grading automorphism. Let $g \in O(V)$ be an orthogonal transformation of V and suppose the Bogoliubov automorphism θ_g of $C(V)$ to be inner, with implementers in $C(M)$ for some $M \in \mathcal{F}(V)$. From above, if θ_g is even (as regards its implementers) then $\theta_g \mid C(M^\perp) = I$ so that $g \mid M^\perp = I$; likewise, if θ_g is odd (as regards implementation) then $\theta_g \mid C(M^\perp) = \gamma$ so that $g \mid M^\perp = -I$. Thus, if θ_g is inner and has implementers in $C(M)$ for some $M \in \mathcal{F}(V)$ then either $M^\perp \subset \ker(g - I)$ (when θ_g is even) or $M^\perp \subset \ker(g + I)$ (when θ_g is odd). Now $M^{\perp\perp} = M$ since M is finite-dimensional and

$$\operatorname{ran}(g \mp I) \subset \ker(g \mp I)^\perp$$

by direct calculation. Consequently, the inclusion $M^\perp \subset \ker(g \mp I)$ entails that

$$M = M^{\perp\perp} \supset \ker(g \mp I)^\perp \supset \operatorname{ran}(g \mp I)$$

and hence that $g \mp I$ has finite rank. We accordingly have the following restriction on orthogonal transformations for which the induced Bogoliubov automorphisms are inner.

Theorem 4.1.4 *Let V be infinite-dimensional and let $g \in O(V)$ be such that θ_g is inner. If θ_g is even then $g - I$ is of finite rank whilst if θ_g is odd then $g + I$ is of finite rank.*   □

Now notice that if $g \in O(V)$ is such that $g \mp I$ has finite rank then $\ker(g \pm I)$ is finite-dimensional: indeed, if v lies in $\ker(g \pm I)$ then $v = \frac{1}{2}(I \mp g)v$ so that in fact

$$\ker(g \pm I) \subset \operatorname{ran}(g \mp I).$$

Accordingly, in deciding whether the Bogoliubov automorphism θ_g of $C(V)$ is inner, our analysis may be divided into four cases, distinguished by the finite-dimensionality of $\operatorname{ran}(g \mp I)$ and the dimensional parity of $\ker(g \pm I)$.

First we claim that if $g - I$ is of finite rank and $\ker(g+I)$ of even dimension, then θ_g is inner and has a unitary implementer in $C^+(\operatorname{ran}(g-I))$. Let us here agree to write $M = \operatorname{ran}(g - I)$ for convenience. Note that $M^\perp = \ker(g - I)$: on the one hand, if $v \in M^\perp$ then $gv \in M^\perp$ so that $gv - v \in M \cap M^\perp = 0$; on the other, if $v \in \ker(g - I)$ and $w \in V$ then

$$(v \mid gw - w) = (g^{-1}v - v \mid w) = (g^{-1}(I - g)v \mid w) = 0.$$

Note also that $g \mid M$ lies not only in $O(M)$ but actually in $SO(M)$ since $\ker(g + I)$ lies in $\operatorname{ran}(g - I) = M$ and is supposed to have even dimension; further, spectral decomposition of $g \mid M$ reveals that M itself is even-dimensional. It now follows from Theorem 4.1.1 that $\theta_{g|M}$ is inner on $C(M)$ with a unitary implementer $u \in C^+(M)$. Being even, u commutes with $C(M^\perp)$ and so implements θ_g on the whole of $C(V)$.

Next we claim that if $g + I$ is of finite rank and $\ker(g - I)$ of odd dimension, then θ_g is inner and has a unitary implementer in $C^-(\operatorname{ran}(g+I))$. Let us here agree to write $M = \operatorname{ran}(g + I)$ for convenience. Note that $M^\perp = \ker(g + I)$: on the one hand, if $v \in M^\perp$ then $gv \in M^\perp$ so that $gv + v \in M \cap M^\perp = 0$; on the other, if $v \in \ker(g + I)$ and $w \in V$ then

$$(v \mid gw + w) = (g^{-1}v + v \mid w) = (g^{-1}(I + g)v \mid w) = 0.$$

Note also that $g \mid M$ lies not simply in $O(M)$ but actually in $SO(M)$ since $g+I$ has kernel $M^\perp$ complementary to M; further, spectral decomposition of $g \mid M$ reveals that M is odd-dimensional, because $\ker(g-I) \subset M$ is supposed to have odd dimension. It now follows from Theorem 4.1.2 that $\theta_{g|M}$ is inner on $C(M)$ with a unitary implementer $u \in C^-(M)$. Being odd, u anticommutes with $C(M^\perp)$ and so implements θ_g on $C(V)$.

We claim finally that θ_g is not inner in the remaining cases. If $g - I$ has finite rank and $\ker(g + I)$ has odd dimension, then $-g^{-1} + I$ has finite rank and $\ker(-g^{-1} - I)$ has odd dimension; if $g + I$ has finite rank and $\ker(g - I)$ has even dimension, then $-g^{-1} - I$ has finite rank and $\ker(-g^{-1} + I)$ has even dimension. In either case, $\theta_{-g^{-1}}$ is inner from what has gone before. The hypothesis that θ_g itself is inner consequently forces upon us the conclusion that the grading automorphism $\gamma = \theta_{-I} = \theta_{-g^{-1}}\theta_g$ is inner and so contradicts Theorem 1.1.20.

We are now able to refine and complete the characterization initiated by Theorem 4.1.4.

Theorem 4.1.5 *Let $g \in O(V)$ where V is infinite-dimensional. The Bogoliubov automorphism θ_g of $C(V)$ is inner if and only if either $g - I$ has finite rank and $\ker(g + I)$ is even-dimensional or $g + I$ has finite rank and $\ker(g - I)$ is odd-dimensional.* □

Our proof of this theorem yielded more information than we chose to record formally. For example, we determined the parity of the implementers; note that this is well-defined, since the infinite-dimensional complex Clifford algebra has scalar centre. Thus: if $g - I$ has finite rank and $\ker(g + I)$ is even-dimensional, then θ_g is implemented by an even unitary in $C^+\big(\operatorname{ran}(g - I)\big)$; if $g + I$ has finite rank and $\ker(g - I)$ is odd-dimensional, then θ_g is implemented by an odd unitary in $C^-\big(\operatorname{ran}(g + I)\big)$.

Incidentally, we observe that the hypothesis in Theorem 4.1.5 that V be infinite-dimensional is not necessary: the conclusions of the theorem apply equally well in case V is of finite dimension. To see this, refer to the spectral decomposition induced by $g \in O(V)$ when V is finite-dimensional: it informs us that $\ker(g - I) \oplus \ker(g + I)$ has the same dimensional parity as V itself. Accordingly, if V is even-dimensional then the dimensions of $\ker(g - I)$ and $\ker(g + I)$ have equal parity, whence the conclusions of Theorem 4.1.5 are those of Theorem 4.1.1 in this case. If V is odd-dimensional then the conditions that $\ker(g + I)$ be even-dimensional and that $\ker(g - I)$ be odd-dimensional are equivalent, characterizing membership of g in $SO(V)$; consequently, the conclusions of Theorem 4.1.5 are those of Theorem 4.1.2 in this case.

This completes our determination of all inner Bogoliubov automorphisms of complex Clifford algebras and puts us in the position of being equipped to introduce the spin groups that were announced in the title of this section. In fact, we shall focus on spin groups over infinite-dimensional real inner product spaces, concluding with some remarks on finite dimensions.

Thus, we shall again let V be an infinite-dimensional real inner product space. Denote by $O_0(V)$ the group of all orthogonal transformations $g \in O(V)$ such that either $g - I$ has finite rank and $\ker(g + I)$ is even-dimensional or $g + I$ has finite rank and $\ker(g - I)$ is odd-dimensional; denote by $SO_0(V)$ the subgroup of $O_0(V)$ comprising all g for which $g - I$ is of finite rank and $\ker(g + I)$ has even dimension. Thus, $O_0(V)$ consists precisely of all orthogonal transformations g of V for which the Bogoliubov automorphism θ_g of $C(V)$ is inner, whilst $SO_0(V)$ consists precisely of all $g \in O(V)$ for which θ_g is inner and has even implementers. Either directly or as a result of these characterizations, it is clear that each of $O_0(V)$ and $SO_0(V)$ is a normal subgroup of $O(V)$.

Let us now consider the special group $SO_0(V)$ a little further. If $g \in SO_0(V)$ then θ_g is an inner automorphism of $C(V)$ having even implementers. These implementers are all proportional, as noted after Theorem 4.1.5; in particular, unitary elements implementing θ_g are proportional by scalars of unit modulus. We shall denote by $\mathrm{Spin}_0^c(V)$ the group of all (even) unitaries implementing the Bogoliubov automorphisms of $C(V)$ induced by elements of $SO_0(V)$. A map

$$\mathrm{Spin}_0^c(V) \to SO_0(V)$$

is well-defined by sending $u \in \mathrm{Spin}_0^c(V)$ to $g \in SO_0(V)$ when θ_g is implemented by u. Quite plainly, this rule defines a surjective group homomorphism having as kernel the group $\mathbb{T}$ of unitary scalars: we have constructed a central short exact sequence of groups

$$1 \to \mathbb{T} \to \mathrm{Spin}_0^c(V) \to SO_0(V) \to 1$$

so that $\mathrm{Spin}_0^c(V)$ is a central extension of $SO_0(V)$ by the circle.

Now, we have defined $\mathrm{Spin}_0^c(V)$ as the group of (necessarily even) unitaries in the complex Clifford algebra $C(V)$ implementing the Bogoliubov automorphisms that are induced by elements of $SO_0(V)$. These automorphisms commute with the main conjugation of $C(V)$ and are thus essentially *real*, whence it is natural to ask for implementers that are likewise essentially real. One way to produce such implementers is to notice that all of the implementers constructed in our determination of the inner Bogoliubov automorphisms actually, lie in the subalgebra of $C(V)$ fixed pointwise by the main conjugation. We opt for an alternative route, via a version of the spinor norm.

Our definition of the spinor norm makes use of the main antiautomorphism α of the complex Clifford algebra, this being the unique antiautomorphism that fixes V pointwise. Let us suppose that $u \in \mathrm{Spin}_0^c(V)$ implements θ_g for $g \in SO_0(V)$. If $v \in V$ then from the defining property $uv = (gv)u$ we deduce that

$$\begin{aligned}
\alpha(u)uv &= \alpha(u)(gv)u \\
&= \alpha\big((gv)u\big)u \\
&= \alpha(uv)u \\
&= v\alpha(u)u.
\end{aligned}$$

Consequently, $\alpha(u)u$ lies in the centre of $C(V)$ and is therefore a scalar; being also unitary, it actually lies in $\mathbb{T}$. Accordingly, a map

$$\nu : \mathrm{Spin}_0^c(V) \to \mathbb{T}$$

is defined by the prescription

$$u \in \mathrm{Spin}_0^c(V) \quad \Rightarrow \quad \nu(u) = \alpha(u)u.$$

This *spinor norm* is in fact a group homomorphism: if $u_1, u_2 \in \mathrm{Spin}_0^c(V)$

then

$$\nu(u_1 u_2) = \alpha(u_1 u_2) u_1 u_2$$
$$= \alpha(u_2) \alpha(u_1) u_1 u_2$$
$$= \alpha(u_1) u_1 \alpha(u_2) u_2$$
$$= \nu(u_1) \nu(u_2)$$

since α is an antiautomorphism and ν takes values in the centre of $C(V)$. Moreover, ν restricts to $\mathbb{T} \subset \mathrm{Spin}_0^c(V)$ as the squaring map: thus, if $\mu \in \mathbb{T}$ and $u \in \mathrm{Spin}_0^c(V)$ then $\nu(\mu u) = \mu^2 \nu(u)$.

We may now define the *spin group* $\mathrm{Spin}_0(V)$ itself to be the kernel of the spinor norm ν in $\mathrm{Spin}_0^c(V)$: thus,

$$\mathrm{Spin}_0(V) = \{ u \in \mathrm{Spin}_0^c(V) : \nu(u) = 1 \}.$$

Since ν restricts to $\mathbb{T}$ as the squaring map, it follows that the kernel of the surjective homomorphism $\mathrm{Spin}_0^c(V) \to SO_0(V)$ intersects $\mathrm{Spin}_0(V)$ in $\{\pm 1\}$ precisely. We therefore have a central short exact sequence of groups

$$1 \to \{\pm 1\} \to \mathrm{Spin}_0(V) \to SO_0(V) \to 1$$

presenting the spin group $\mathrm{Spin}_0(V)$ as a double cover of $SO_0(V)$.

We should check on the reality of our spin group. If $u \in \mathrm{Spin}_0(V)$ then the unitary nature of u and the definition of the main involution on $C(V)$ imply that

$$u^{-1} = u^* = \overline{\alpha(u)};$$

as a consequence,

$$1 = \nu(u) = \alpha(u) u = \overline{u}^{-1} u$$

and therefore $\overline{u} = u$. Thus, the elements of $\mathrm{Spin}_0(V)$ are fixed by the main conjugation on $C(V)$ as promised.

The foregoing construction of the spin group extends easily to yield what is known as the pin group. Explicitly, we denote by $\mathrm{Pin}_0^c(V)$ the group of all unitaries in $C(V)$ that implement Bogoliubov automorphisms arising from elements of $O_0(V)$ and find that there is a central short exact sequence of groups

$$1 \to \mathbb{T} \to \mathrm{Pin}_0^c(V) \to O_0(V) \to 1.$$

The prescription

$$\nu : \mathrm{Pin}_0^c(V) \to \mathbb{T} : u \mapsto \alpha(u) u$$

well-defines a unitary character of $\mathrm{Pin}_0^c(V)$ whose kernel $\mathrm{Pin}_0(V)$ fits into a central short exact sequence of groups

$$1 \to \{\pm 1\} \to \mathrm{Pin}_0(V) \to O_0(V) \to 1.$$

Again, $\mathrm{Pin}_0(V)$ is the real part of $\mathrm{Pin}_0^c(V)$.

We draw our account of the purely algebraic spin groups to a close by outlining the finite-dimensional picture. When V is even-dimensional, the changes are essentially notational: we may drop the subscript zero to obtain groups Spin (V) and Pin (V) by the same procedures. When V is odd-dimensional, the changes are more substantial. If $g \in SO(V)$ then θ_g is implemented by even unitary elements of $C(V)$ as in Theorem 4.1.2; these even unitary implementers are proportional, since even elements in the centre of $C(V)$ are scalar. Thus we obtain a central short exact sequence of groups

$$1 \rightarrow \mathbb{T} \rightarrow \text{Spin}^c(V) \rightarrow SO(V) \rightarrow 1.$$

Moreover, if $u \in \text{Spin}^c(V)$ then $\nu(u) = \alpha(u)u$ is again a unitary in the centre of $C(V)$; being even, it is a scalar. Thus we obtain a unitary character $\nu : \text{Spin}^c(V) \rightarrow \mathbb{T}$ whose kernel Spin (V) is a double cover of $SO(V)$. However, there can be no analogue of the full pin group as a group of implementers for the Bogoliubov automorphisms, since Theorem 4.1.2 warns us that if $g \in O^-(V)$ then θ_g is not inner.

4.2 C^* spin groups

In this section, we are occupied with the construction of a spin group that resides within the C^* Clifford algebra $C[V]$ when V is a real Hilbert space of infinite dimension. Our main problem is to determine precise necessary and sufficient conditions on the orthogonal transformation $g \in O(V)$ in order that the Bogoliubov automorphism θ_g of $C[V]$ is inner. It turns out that θ_g is inner as an automorphism of $C[V]$ precisely when either $g - I$ is of trace class and the kernel of $g + I$ is even-dimensional or $g + I$ is of trace class and the kernel of $g - I$ is odd-dimensional. Not being purely algebraic, this result has a somewhat more involved proof than that of the analogous result for the plain complex Clifford algebra given in the preceding section; however, some of the basic steps in that proof can be recognized here.

First of all, we note that if $g \in O(V)$ is such that the Bogoliubov automorphism θ_g of $C[V]$ is inner, then any invertible element $u \in C[V]$ implementing θ_g has a definite parity. Indeed, it may be seen (as for the complex Clifford algebra) that $\gamma(u)$ also implements θ_g and is hence (by Theorem 1.2.10) proportional to u itself; the fact that $\gamma^2 = I$ again forces the scalar of proportionality to be ± 1. Thus, either $\gamma(u) = u$ or $\gamma(u) = -u$; we may speak of g as being *even* in the former case and *odd* in the latter, since implementers for θ_g are proportional and hence share the same parity. We note also that if θ_g is inner then it has a unitary implementer. Indeed, if the unit u implements θ_g then the unit $(u^{-1})^* = (u^*)^{-1}$ does also, since

$$(u^*)^{-1}au^* = (ua^*u^{-1})^*$$
$$= \left(\theta_g(a^*)\right)^*$$
$$= \theta_g(a)$$

whenever $a \in C[V]$; it follows that the positive u^*u is a scalar, which may be normalized to 1 by scaling.

Our next step is to show that if $g \in O(V)$ and $\theta_g \in \operatorname{Aut} C[V]$ is inner, then either $g - I$ or $g + I$ is at least compact. In this, we are aided by the following estimate concerning the adjoint representation of $C[V]$.

Theorem 4.2.1 *If $u, w \in C[V]$ are such that u is invertible and*

$$3\|u - w\| \le \min\{\|u\|, \|u^{-1}\|^{-1}\}$$

then w is invertible and

$$\|\operatorname{Ad}_u - \operatorname{Ad}_w\| \le \|u^{-1}\| \left(1 + 2\|u\|\,\|u^{-1}\|\right) \|u - w\|.$$

Proof Observe first that the inequality $\|u - w\| < \|u^{-1}\|^{-1}$ ensures that w is invertible: indeed,

$$w^{-1} = u^{-1} \sum_{n \ge 0} (1 - wu^{-1})^n$$

so that

$$\|w^{-1}\| \le \frac{\|u^{-1}\|}{1 - \|1 - wu^{-1}\|}$$
$$\le \frac{\|u^{-1}\|}{1 - \|u - w\|\,\|u^{-1}\|}.$$

Since also

$$\|w\| \le \|u\| + \|u - w\|$$

it follows that

$$\|w\|\,\|w^{-1}\| \le \frac{\|u\| + \|u - w\|}{\|u^{-1}\|^{-1} - \|u - w\|}.$$

From $3\|u - w\| \le \|u\|$ we deduce that

$$\|u\| + \|u - w\| \le \frac{4}{3}\|u\|$$

and from $3\|u - w\| \le \|u^{-1}\|^{-1}$ we deduce that

$$\|u^{-1}\|^{-1} - \|u - w\| \ge \frac{2}{3}\|u^{-1}\|^{-1}.$$

Consequently,

$$\|w\|\,\|w^{-1}\| \le 2\|u\|\,\|u^{-1}\|.$$

Now, if $a \in C[V]$ then

$$\|\mathrm{Ad}_u a - \mathrm{Ad}_w a\| = \|uau^{-1} - waw^{-1}\|$$
$$= \| (u - w)au^{-1} + wa(u^{-1} - w^{-1}) \|$$
$$\leq \|u - w\| \, \|a\| \, \|u^{-1}\| + \|w\| \, \|a\| \, \|u^{-1}\| \, \|u - w\| \, \|w^{-1}\|$$
$$= \|u^{-1}\| \, (1 + \|w\| \, \|w^{-1}\|) \, \|u - w\| \, \|a\|.$$

We conclude that

$$\|\mathrm{Ad}_u - \mathrm{Ad}_w\| \leq \|u^{-1}\| \, (1 + 2\|u\| \, \|u^{-1}\|) \, \|u - w\|$$

as was to be shown. $\square$

Notice here that if the element $u \in C[V]$ happens to be unitary, then matters simplify: in this case,

$$3\|u - w\| \leq 1 \quad \Rightarrow \quad \|\mathrm{Ad}_u - \mathrm{Ad}_w\| \leq 3\|u - w\|.$$

In fact, this special case of the theorem is all that we shall require in what follows.

Now, let $g \in O(V)$ be such that θ_g is an inner automorphism of $C[V]$ and suppose first that g is *even*; indeed, let θ_g be implemented by the even unitary $u \in C^+[V]$. In order to show that $g - I$ is compact in this case, we shall show that it maps weakly convergent sequences to strongly convergent ones. As preparation for this, let $\epsilon > 0$ be given and assume $\epsilon \leq \frac{1}{3}$ without loss. Since the even complex Clifford algebra is uniformly dense in the even C^* Clifford algebra, there exists $w \in C^+(V)$ such that $\|u - w\| \leq \epsilon$. The simplified version of Theorem 4.2.1 now implies that w is invertible and that $\|\mathrm{Ad}_u - \mathrm{Ad}_w\| \leq 3\epsilon$. Of course, w lies in $C^+(M)$ for some finite-dimensional subspace $M \in \mathcal{F}(V)$ of V, by Theorem 1.1.13. Note from the Clifford relations that, being even, w commutes with each element of $C(M^\perp)$.

So, let $(v_n : n > 0)$ be a sequence in V that is weakly convergent to zero; we wish to show that $(gv_n - v_n : n > 0)$ is strongly convergent to zero. For each $n > 0$ let us put $v_n = x_n + y_n$ with $x_n \in M$ and $y_n \in M^\perp$. The finite-dimensionality of M implies that $(x_n : n > 0)$ is not only weakly convergent to zero but actually strongly convergent to zero. The uniform boundedness principle implies that the weakly convergent sequence $(v_n : n > 0)$ is strongly bounded and hence so is $(y_n : n > 0)$ with the same bound. Since u implements θ_g and since w commutes with elements of $M^\perp$ we have that

$$gv_n - v_n = uv_n u^{-1} - v_n$$
$$= \mathrm{Ad}_u v_n - v_n$$
$$= \mathrm{Ad}_u x_n - x_n + (\mathrm{Ad}_u - \mathrm{Ad}_w)y_n$$

whence
$$\|gv_n - v_n\| \leq \| \operatorname{Ad}_u x_n - x_n \| + \| (\operatorname{Ad}_u - \operatorname{Ad}_w) y_n \|$$
$$\leq (\| \operatorname{Ad}_u \| + 1) \|x_n\| + \| \operatorname{Ad}_u - \operatorname{Ad}_w \| \|y_n\|$$
$$\leq (\| \operatorname{Ad}_u \| + 1) \|x_n\| + 3\epsilon \|y_n\|.$$

Since $\epsilon \in (0, \frac{1}{3}]$ is arbitrary, since $(x_n : n > 0)$ is strongly convergent to zero and since $(y_n : n > 0)$ is strongly bounded by a constant depending on $(v_n : n > 0)$ only, we conclude that $\|gv_n - v_n\| \to 0$ as $n \to \infty$.

Supposing instead that g is *odd*, a similar argument establishes that the operator $g + I$ is compact. We remark only that in this case, the element chosen to approximate an implementer of θ_g will lie in the odd space $C^-(M)$ for some $M \in \mathcal{F}(V)$ and will therefore anticommute with elements of $M^{\perp}$.

In summary, we have completed the next step in our determination of all inner Bogoliubov automorphisms of the C^* Clifford algebra.

Theorem 4.2.2 *Let $g \in O(V)$ be such that the Bogoliubov automorphism θ_g of $C[V]$ is inner. If g is even then $g - I$ is compact; if g is odd then $g + I$ is compact.*   □

Recall here that by the *parity* of g we mean the parity of any implementer for θ_g.

We now confine our attention to $g \in O(V)$ such that either $g - I$ or $g + I$ is compact. According to the spectral theorem for compact operators, there is an orthogonal decomposition
$$V = W^+ \oplus W \oplus W^-$$
in which
$$W^{\pm} = \ker (g \mp I)$$
and in which W further decomposes orthogonally as
$$W = \bigoplus_{n>0} W_n$$
where if $n > 0$ then W_n is a plane on which g acts by rotation through an angle φ_n in the range $0 < |\varphi_n| < \pi$. For each $n > 0$ we choose an orthonormal basis $\{x_n, y_n\}$ of W_n relative to which $g_n = g \mid W_n$ has matrix
$$[g_n] = \begin{bmatrix} \cos \varphi_n & -\sin \varphi_n \\ \sin \varphi_n & \cos \varphi_n \end{bmatrix}$$
so that
$$g \cdot x_n = \cos \varphi_n \cdot x_n + \sin \varphi_n \cdot y_n$$
$$g \cdot y_n = -\sin \varphi_n \cdot x_n + \cos \varphi_n \cdot y_n.$$

In particular, note that if $g - I$ is compact then its eigenspace W^- is necessarily finite-dimensional; if additionally W is infinite-dimensional, then $\varphi_n \to 0$ as $n \to \infty$. Note also the following formula for the trace norm of the operator $g - I$ in general.

Theorem 4.2.3 *The trace norm of $g - I$ is given by*

$$\|g - I\|_1 = 4 \sum_{n>0} |\sin \tfrac{1}{2}\varphi_n| + 2 \dim W^-.$$

Proof If $n > 0$ then a routine computation yields

$$|g_n - I|^2 = (g_n - I)^*(g_n - I) = 2(1 - \cos\varphi_n)I$$

whence

$$|g_n - I| = \sqrt{2(1 - \cos\varphi_n)}\, I = 2|\sin \tfrac{1}{2}\varphi_n|\, I$$

so that, since W_n is a real plane,

$$\|g_n - I\|_1 = \operatorname{Tr}|g_n - I| = 4|\sin \tfrac{1}{2}\varphi_n|.$$

Once the facts $(g - I) \mid W^+ = 0$ and $(g - I) \mid W^- = -2I$ are taken into account, the theorem follows upon summation. $\qquad\square$

Having been given $g \in O(V)$ with the property that either $g - I$ or $g + I$ is compact, let us now continue our analysis. This splits into four cases distinguished by the dimensional nature of $W^+ = \ker(g - I)$ and $W^- = \ker(g + I)$: that is, distinguished by whether these kernels are odd-dimensional or otherwise. Of course, if $g \mp I$ is compact then its eigenspace $\ker(g \pm I)$ is finite-dimensional, but $\ker(g \mp I)$ might be infinite-dimensional.

The *first case* in our analysis involves by far the most effort: that in which neither W^+ nor W^- is odd-dimensional. In this case we equip W^+ and W^- with unitary structures, which we extend to a unitary structure J on the whole of V by stipulating that $Jx_n = y_n$ and $Jy_n = -x_n$ whenever $n > 0$. Notice that the given orthogonal transformation $g \in O(V)$ commutes with J and so in fact defines a unitary transformation $g \in U(V_J)$ on V made complex Hilbert via J. In particular, notice that if $n > 0$ then g acts on the complex line W_n as multiplication by $\exp(i\varphi_n)$.

The introduction of a unitary structure places at our disposal the theory of Fock representations. In fact, let π_J be the Fock representation of $C[V]$ on the Fock space $\mathbb{H}_J$ induced by the unitary structure J introduced above. The transformation $g \in U(V_J)$ being unitary, Theorem 3.2.5 tells us that the Bogoliubov automorphism θ_g of $C[V]$ is implemented in π_J by the unitary operator $\bigwedge(g) = \bigwedge_g$ on $\mathbb{H}_J$: thus,

$$a \in C[V] \quad \Rightarrow \quad \pi_J(\theta_g a) = \bigwedge{}_g \circ \pi_J(a) \circ \bigwedge{}_g^*.$$

Moreover, since g commutes with $-I$, the implementing operator $\bigwedge(g)$ commutes with the grading operator $\bigwedge(-I) = \Gamma_J$ on Fock space. As we shall see, these facts will prove invaluable to our analysis of this first case.

Before embarking upon the analysis proper, it is convenient to set up a little more notation. For each $n > 0$ let

$$Z_n = W_1 \oplus \cdots \oplus W_n \oplus W^-,$$

let $h_n = g \mid Z_n$ and let $k_n = g \mid Z_n^\perp$. It follows that Fock space $\mathbb{H}_J = \bigwedge[V_J]$ decomposes as the tensor product

$$\mathbb{H}_J = \bigwedge[Z_n^\perp] \otimes \bigwedge[Z_n]$$

and that

$$\bigwedge(g) = \bigwedge(k_n) \otimes \bigwedge(h_n)$$

accordingly. If we also put $h = g \mid W \oplus W^-$ then

$$\bigwedge(g) = I \otimes \bigwedge(h)$$

corresponding to the decomposition

$$\mathbb{H}_J = \bigwedge[W^+] \otimes \bigwedge[W \oplus W^-].$$

Let us write $\mathcal{S}$ for the empty index $\emptyset$ together with the collection of all strictly increasing integer multiindices $S = (s_1, \ldots, s_l)$ with $0 < s_1 < \ldots < s_l$ and $l > 0$. The vectors $\{x_n : n > 0\}$ form a complete orthonormal system for W as a complex Hilbert space. Consequently, $\{x_S : S \in \mathcal{S}\}$ is a complete orthonormal system for $\bigwedge[W]$ where $x_\emptyset = \Omega_J \in \bigwedge[W]$ and where $x_S = x_{s_1} \wedge \ldots \wedge x_{s_l}$ if $S = (s_1, \ldots, s_l)$. If $n \geq 0$ then let us also write $\mathcal{S}_n$ for the subset of $\mathcal{S}$ comprising all multiindices having entries strictly larger than n.

Now, recall that if $s > 0$ then g acts as multiplication by $\exp(i\varphi_s)$ on the complex line W_s. Upon taking exterior products, there results the following fact.

Theorem 4.2.4   *If $S = (s_1, \ldots, s_l) \in \mathcal{S}$ then $x_S \in \bigwedge[W]$ is an eigenvector with eigenvalue*

$$\exp\left\{i \sum_{j=1}^{l} \varphi_{s_j}\right\}$$

for $\bigwedge(g) = I \otimes \bigwedge(h)$ and with eigenvalue

$$\exp\left\{i \sum_{s_j \leq n} \varphi_{s_j}\right\}$$

for $I \otimes \bigwedge(h_n)$ *whenever* $n > 0$. $\qquad\square$

As a corollary of this result, we are able to compute the operator norm of $I \otimes \bigwedge(h) - I \otimes \bigwedge(h_n)$ when $n > 0$. Indeed, if $S = (s_1, \ldots, s_l) \in \mathcal{S}$ then

$$\big(I \otimes \bigwedge(h)\big)x_S = \exp\Big\{i \sum_{j=1}^{l} \varphi_{s_j}\Big\} x_S$$

and

$$\big(I \otimes \bigwedge(h_n)\big)x_S = \exp\Big\{i \sum_{s_j \leq n} \varphi_{s_j}\Big\} x_S$$

so that x_S is an eigenvector for the operator $I \otimes \bigwedge(h) - I \otimes \bigwedge(h_n)$ with eigenvalue of modulus

$$\Big| \exp\Big\{i \sum_{s_j > n} \varphi_{s_j}\Big\} - 1\Big|.$$

Since $\{x_S : S \in \mathcal{S}\}$ is a complete orthonormal system for $\bigwedge[W]$ and since $g \mid W^{\pm} = \pm I$, spectral theory now yields the following result.

Theorem 4.2.5 *If $n > 0$ then the difference $\bigwedge(g) - I \otimes \bigwedge(h_n)$ has operator norm given by the supremum of the numbers*

$$\Big| \exp\Big\{i \sum_{s \in S} \varphi_s\Big\} - 1 \Big|$$

as S ranges over $\mathcal{S}_n$. $\qquad\square$

Now, on to the detailed analysis of our first case. If we suppose that $\theta_g \in \operatorname{Aut} C[V]$ is inner and implemented by the unitary $u \in C[V]$ then the unitary operator $\pi_J(u)$ on $\mathbb{H}_J$ implements θ_g in π_J and is therefore a scalar multiple of $\bigwedge(g)$ since π_J is irreducible. As $\bigwedge(g)$ commutes with Γ_J it follows that $\pi_J(u)$ does also, whence Theorem 2.5.1 implies that $u = \gamma(u)$ is even since π_J is faithful. Theorem 4.2.2 now informs us that in this case, the operator $g - I$ is actually compact; its eigenspace W^- is consequently finite-dimensional.

Our first case thus has the following hypotheses on $g \in O(V)$: that $g - I$ is compact, that W^+ is other than odd-dimensional and that W^- is even-dimensional. The remainder of our analysis in this case has the following strategy: to show that each of the circumstances, that θ_g be inner and that $g - I$ be trace class, is equivalent to the circumstance that $I \otimes \bigwedge(h_n) \to \bigwedge(g)$ as $n \to \infty$.

Suppose θ_g to be inner; indeed, let it be implemented by the even unitary $u \in C^+[V]$. After multiplication by a scalar of unit modulus, we may arrange that $\pi_J(u) = \bigwedge(g)$. Since u implements θ_g and since

$g \mid W^+ = I$, it follows that u commutes with each element of W^+. Being even, u therefore lies in the C^* Clifford algebra $C[(W^+)^\perp] = C[W \oplus W^-]$ according to Theorem 1.2.20. The fact that the union

$$\bigcup \{C(Z_n) : n > 0\}$$

is dense in $C[W \oplus W^-]$ allows us to choose for each $n > 0$ an element $z_n \in C(Z_n)$ such that $z_n \to u$ as $n \to \infty$. Plainly, we may assume that if $n > 0$ then z_n is invertible; in fact, we may assume that if $n > 0$ then z_n is unitary. To see this, note that $C(Z_n)$ is a C^* algebra since Z_n is finite-dimensional, and let z_n have polar decomposition $z_n = u_n|z_n|$ where $u_n \in C(Z_n)$ is unitary. Since $z_n \to u$ it follows that $|z_n|^2 = z_n^* z_n \to u^* u = \mathbf{1}$ and hence $|z_n| \to \mathbf{1}$ by continuity. Since also

$$\|u - u_n\| \leq \|u - z_n\| + \|u_n|z_n| - u_n\|$$
$$\leq \|u - z_n\| + \|\,|z_n| - \mathbf{1}\|$$

it follows that $u_n \to u$. Thus: we may assume unitaries $u_n \in C(Z_n)$ chosen so that $u_n \to u$. Notice that the unitary operator $\pi_J(u_n)$ on $\mathbb{H}_J$ has the form $I \otimes U_n$ on $\bigwedge[Z_n^\perp] \otimes \bigwedge[Z_n]$ for some unitary U_n on $\bigwedge[Z_n]$ and that $I \otimes U_n \to \bigwedge(g)$. Consequently, the operators $I \otimes U_n$ and $\bigwedge(g) = \bigwedge(k_n) \otimes \bigwedge(h_n)$ come arbitrarily close for n sufficiently large. We claim that this closeness to $\bigwedge(g)$ is improved by taking $I \otimes \bigwedge(h_n)$ in place of $I \otimes U_n$. This claim will be justified by the following technical result.

Theorem 4.2.6 *Let X and Y be unitary operators on the complex Hilbert spaces $\mathbb{H}$ and $\mathbb{K}$ respectively. If U is a unitary operator on $\mathbb{H}$ then there exists $\mu \in \mathbb{T}$ such that*

$$\|I \otimes \mu X - Y \otimes X\| \leq \|I \otimes U - Y \otimes X\|.$$

Proof On the one hand, since X and Y are unitary operators we have the equality

$$\| I \otimes U - Y \otimes X\| = \| Y^{-1} \otimes (UX^{-1}) - I \otimes I \|;$$

on the other, we have similarly

$$\| I \otimes \mu X - Y \otimes X \| = \| \mu Y^{-1} - I \|.$$

We are thus reduced to showing that if X and Y are unitary operators on $\mathbb{H}$ and $\mathbb{K}$ respectively then there exists $\mu \in \mathbb{T}$ such that

$$\|\mu Y - I\| \leq \|Y \otimes X - I \otimes I\|.$$

To see this, note first that if $\xi \in \sigma(X)$ and $\eta \in \sigma(Y)$ then $\eta\,\xi$ lies in the spectrum $\sigma(Y \otimes X)$ of $Y \otimes X$. Now choose $\mu \in \sigma(X)$ so as to minimize

$\|\mu Y - I\|$; then

$$\|\mu Y - I\| = \sup\{|\mu\nu - 1| : \nu \in \sigma(Y)\}$$
$$\leq \sup\{|\lambda - 1| : \lambda \in \sigma(Y \otimes X)\}$$
$$= \|Y \otimes X - I \otimes I\|$$

and we are finished. $\qquad\qquad\qquad\qquad\qquad\qquad\qquad\qquad\qquad\square$

On the strength of this result, for each $n > 0$ we may choose $\mu_n \in \mathbb{T}$ so that

$$\|I \otimes \mu_n \bigwedge(h_n) - \bigwedge(g)\| \leq \|I \otimes U_n - \bigwedge(g)\|$$

whence $I \otimes \mu_n \bigwedge(h_n) \to \bigwedge(g)$. Consequently,

$$\mu_n \Omega_J = (I \otimes \mu_n \bigwedge(h_n))\Omega_J \to \bigwedge(g)\Omega_J = \Omega_J$$

so that $\mu_n \to 1$ and hence in fact $I \otimes \bigwedge(h_n) \to \bigwedge(g)$.

We claim that $g - I$ is of trace class. If not, then Theorem 4.2.3 and the finite-dimensionality of W^- imply that the series $\sum_{n>0} |\sin \frac{1}{2}\varphi_n|$ diverges, whence so does the series $\sum_{n>0} |\varphi_n|$ by the limit comparison test, noting that $\varphi_n \to 0$. As a consequence, the series $\sum_{n>0} \varphi_n$ is such that either its subseries of positive terms or its subseries of negative terms diverges; with no essential loss, assume

$$\sum \{\varphi_n : \varphi_n > 0\} = \infty.$$

Since the sequence $(\varphi_n : n > 0)$ converges to zero, if $m > 0$ is given then we may choose $s_l > \ldots > s_1 > m$ so that

$$\left| \sum_{j=1}^{l} \varphi_{s_j} - \pi \right| \leq \frac{\pi}{2}$$

and therefore

$$\left| \exp\left\{ i \sum_{j=1}^{l} \varphi_{s_j} \right\} - 1 \right| \geq \sqrt{2}.$$

Theorem 4.2.4 tells us that if $n > s_l$ then $I \otimes \bigwedge(h_n) - I \otimes \bigwedge(h_m)$ has

$$\exp\left\{ i \sum_{j=1}^{l} \varphi_{s_j} \right\} - 1$$

as an eigenvalue, so it follows that

$$\|I \otimes \bigwedge(h_n) - I \otimes \bigwedge(h_m)\| \geq \sqrt{2}.$$

In particular, the sequence $(I \otimes \bigwedge(h_n) : n > 0)$ cannot possibly converge, contrary to the finding of the previous paragraph. Thus, $g - I$ is indeed of trace class.

Still within our first case, let us suppose conversely that $g - I$ is of trace class. If $\epsilon \in (0, \frac{\pi}{2}]$ then there exists $n_\epsilon > 0$ such that $\sum_{j>n_\epsilon} |\varphi_j| \leq \epsilon$ on account of Theorem 4.2.4 and the limit comparison test. Thus, if $S \in \mathcal{S}_{n_\epsilon}$ then

$$\left| \sum_{s \in S} \varphi_s \right| \leq \sum_{s \in S} |\varphi_s| \leq \epsilon$$

and so

$$\left| \exp\left\{ i \sum_{s \in S} \varphi_s \right\} - 1 \right| \leq \epsilon.$$

From Theorem 4.2.5 it now follows that

$$n \geq n_\epsilon \quad \Rightarrow \quad \left\| I \otimes \bigwedge(h_n) - \bigwedge(g) \right\| \leq \epsilon$$

whence the sequence $(I \otimes \bigwedge(h_n) : n > 0)$ converges to $\bigwedge(g)$ in operator norm. If $n > 0$ then the Fock representation π_J sets up an isomorphism between $C(Z_n)$ and $B(\bigwedge[Z_n])$ since Z_n is even-dimensional. As $\bigwedge(h_n)$ is a unitary operator on $\bigwedge[Z_n]$ with which the grading operator $\bigwedge(-I)$ commutes, it follows that there exists a unique even unitary $u_n \in C^+(Z_n)$ such that $\pi_J(u_n) = I \otimes \bigwedge(h_n)$. Recalling that π_J is isometric, it follows from $I \otimes \bigwedge(h_n) \to \bigwedge(g)$ that there exists an even unitary $u \in C^+[W \oplus W^-]$ such that $u_n \to u$ and $\pi_J(u) = \bigwedge(g)$. Now, if $a \in C[V]$ then

$$\begin{aligned} \pi_J(\theta_g a) &= \bigwedge(g) \pi_J(a) \bigwedge(g)^* \\ &= \pi_J(u) \pi_J(a) \pi_J(u)^* \\ &= \pi_J(uau^*) \end{aligned}$$

whence

$$\theta_g(a) = uau^*$$

since π_J is faithful. Thus $\theta_g \in \operatorname{Aut} C[V]$ is inner, with $u \in C^+[V]$ as an even unitary implementer.

At last, our analysis of the first case is complete: if neither W^+ nor W^- is odd-dimensional, then θ_g is inner precisely when $g - I$ is trace-class; moreover, implementers are invariably even in this case. Note also that θ_g being inner and $g - I$ being trace-class are each equivalent to $\bigwedge(g)$ lying in the range of π_J; here, J is the specific unitary structure on V introduced in the course of our analysis.

As the *second case* in our analysis, we take that in which both W^+ and W^- are odd-dimensional. Choose unit vectors $x \in W^+$ and $y \in W^-$ and let M be the plane that they span. Equip $M^\perp$ with a unitary structure J by extending any unitary structures on the orthocomplements $W^+ \ominus \mathbb{R}x$ and $W^- \ominus \mathbb{R}y$ according to $Jx_n = y_n$ when $n > 0$; extend J to a unitary structure on the whole of V by requiring that $Jx = y$ and of course

$Jy = -x$. The orthogonal transformation $h \in O(V)$ determined by the conditions $h \mid M^{\perp} = g \mid M^{\perp}$ and $h \mid M = I$ now lies in $U(V_J)$ and satisfies

$$\ker(h - I) = W^{+} \oplus \mathbb{R}y$$
$$\ker(h + I) = W^{-} \ominus \mathbb{R}y$$

with either $h - I$ or $h + I$ compact. As $yyy = y$ and $yvy = -v$ if v and y are orthogonal, it follows that the unitary operator $\bigwedge(-h) \circ \pi_J(y)$ implements θ_g in the Fock representation π_J of $C[V]$ on $\mathbb{H}_J = \bigwedge[V_J]$. Since π_J is irreducible and faithful, we deduce that θ_g is inner if and only if $\bigwedge(-h) \in \pi_J(C[V])$ in view of the fact that $\pi_J(y)$ lies in $\pi_J(C[V])$ already. The argument of the first case now applies: $\bigwedge(-h) \in \pi_J(C[V])$ if and only if $(-h) - I$ is trace-class. Since $\operatorname{ran}(g - h) = \mathbb{R}y$ is finite-dimensional, $(-h) - I$ is trace-class if and only if $g + I$ is trace-class. Thus, if both W^{+} and W^{-} are odd-dimensional then θ_g is inner precisely when $g + I$ is trace-class; moreover, implementers are invariably odd in this case.

For the *third case*, let the dimension of W^{+} be odd and that of W^{-} not. Choose a unit vector $l \in W^{+}$ and denote by L its linear span. Define a unitary structure J on $L^{\perp}$ by extending any unitary structures on $W^{+} \ominus L$ and W^{-} according to $Jx_n = y_n$ when $n > 0$. The restriction $h = g \mid L^{\perp}$ then lies in $U(L^{\perp}{}_J)$ and satisfies

$$\ker(h - I) = W^{+} \ominus L$$
$$\ker(h + I) = W^{-}$$

with either $h - I$ or $h + I$ compact. Now extend the Fock representation π_J of $C[L^{\perp}]$ on $\bigwedge[L^{\perp}{}_J]$ to a representation

$$\pi : C[V] \to B\big(\bigwedge[L^{\perp}{}_J]\big)$$

by requiring that $\pi(l) = \Gamma_J$ along with $\pi \mid L^{\perp} = \pi_J$. Since h and J commute, $\bigwedge(h)$ certainly implements θ_h in the Fock representation π_J of $C[L^{\perp}]$. Since $\bigwedge(h)$ and $\bigwedge(-I) = \Gamma_J = \pi(l)$ commute, it follows that $\bigwedge(h)$ actually implements θ_g in the representation π of $C[V]$. Our analysis of this case now bifurcates into two subcases divided by parity.

Let θ_g be inner and implemented by the *even* unitary $u \in C^{+}[V]$. In particular, $ulu^{-1} = \theta_g l = l$ so that $lul = u$, as a consequence of which u actually lies in $C^{+}[L^{\perp}]$ according to Theorem 1.2.16. Now the unitary operators $\bigwedge(h)$ and $\pi(u) = \pi_J(u) \in \pi_J(C[L^{\perp}])$ both implement θ_g and are hence proportional by irreducibility, so $\bigwedge(h) \in \pi_J(C[L^{\perp}])$. By means of our first case, we deduce that $h - I$ is trace-class, whence $g - I$ itself is trace-class. Conversely if $g - I$ is trace-class then so is $h - I$;

our first case now tells us that θ_h is inner with an even implementer, whence θ_g itself is inner.

Let θ_g be inner and implemented by the *odd* unitary $u \in C^-[V]$. Again l and u commute, thus $l(ul)l = lu = ul$ and so $ul \in C^+[L^\perp]$ by Theorem 1.2.16. Let $k \in O(V)$ be reflection in the hyperplane perpendicular to l so that $k \mid L^\perp = I$ and θ_{-k} is inner with l as implementer. Now ul implements θ_{-gk} and hence implements θ_{-h} since $gk \mid L^\perp = h$. By means of our first case we deduce that $-h - I$ is trace-class, whence $g + I$ is trace-class. Conversely, if $g + I$ is trace-class then so is $-h - I$; our first case now tells us that θ_{-h} is inner and even, thus θ_{-gk} is inner and so $\theta_g = \theta_{-gk}\theta_{-k}$ is inner.

Note that in this third case, implementers come in each parity.

Our *fourth and final case* to consider is that in which W^- is odd-dimensional and W^+ is otherwise. Here we choose a unit vector $l \in W^-$ having L as its linear span. Extend any unitary structures on W^+ and $W^- \ominus L$ to define a unitary structure J on $L^\perp$ with $Jx_n = y_n$ for $n > 0$. The restriction $h = g \mid L^\perp$ then lies in $U(L^\perp {}_J)$ and satisfies

$$\ker(h - I) = W^+$$
$$\ker(h + I) = W^- \ominus L$$

with either $h - I$ or $h + I$ compact. We again extend the Fock representation π_J of $C[L^\perp]$ on $\bigwedge[L^\perp {}_J]$ to a representation

$$\pi : C[V] \to B\left(\bigwedge[L^\perp {}_J]\right)$$

by demanding not only that $\pi \mid L^\perp = \pi_J$ but also that $\pi(l) = \Gamma_J$. Notice that the unitary operator $\bigwedge(h)$ implements θ_h in the Fock representation π_J. If θ_g is inner with the unitary $u \in C[V]$ as implementer, then the unitary operator $\pi(u)$ on $\bigwedge[L^\perp {}_J]$ implements θ_g in π and hence implements θ_h in π_J. The irreducibility of π_J now forces $\bigwedge(h)$ and $\pi(u)$ to be proportional. However: on the one hand, $\bigwedge(h)$ and $\bigwedge(-I) = \Gamma_J$ commute since h and $-I$ commute; on the other, $\pi(u)$ anticommutes with $\pi(l) = \Gamma_J$ since

$$-\pi(l) = \pi(\theta_g l) = \pi(u)\pi(l)\pi(u)^*.$$

This contradiction precludes the existence of u. Thus, θ_g is never inner in this final case.

At long last, our determination of all inner Bogoliubov automorphisms for the C^* Clifford algebra is complete. We record the essentials of the determination in the following form.

Theorem 4.2.7 *If $g \in O(V)$ then the Bogoliubov automorphism θ_g of $C[V]$ is: inner and even precisely when $g - I$ is trace-class and* ker $(g + I)$ *is even-dimensional; inner and odd precisely when* $g + I$ *is trace-class and* ker $(g - I)$ *is odd-dimensional.* □

Our proof brought to light rather more information than we chose to record; we are content to leave this extra information where it lies.

We are naturally led to denote by $O_1(V)$ the set of all orthogonal transformations $g \in O(V)$ such that either $g - I$ is trace-class and ker $(g+I)$ is even-dimensional or $g+I$ is trace-class and ker $(g-I)$ is odd-dimensional. Thus, $O_1(V)$ is the normal subgroup of $O(V)$ comprising all orthogonal transformations of V for which the induced Bogoliubov automorphism of $C[V]$ is inner. We are also led to denote by $SO_1(V)$ the set of all $g \in O(V)$ such that $g-I$ is trace-class and ker $(g+I)$ is even-dimensional. Thus, $SO_1(V)$ is the normal subgroup of $O(V)$ comprising all orthogonal transformations of V for which the induced Bogoliubov automorphism of $C[V]$ is inner and even.

Now we denote by $\mathrm{Spin}\,_1^c(V)$ the group of all (even) unitary elements of $C[V]$ that implement the Bogoliubov automorphisms of $C[V]$ induced by elements of $SO_1(V)$. A group homomorphism

$$\mathrm{Spin}\,_1^c(V) \to SO_1(V)$$

is well-defined by assigning $g \in SO_1(V)$ to $u \in \mathrm{Spin}\,_1^c(V)$ in case u implements θ_g. The kernel of this homomorphism is the group $\mathbb{T}$ of unitary scalars, since the C^* Clifford algebra is central. Thus, we arrive at a central short exact sequence of groups

$$1 \to \mathbb{T} \to \mathrm{Spin}\,_1^c(V) \to SO_1(V) \to 1$$

and $\mathrm{Spin}\,_1^c(V)$ is a central extension of $SO_1(V)$ by the circle.

Note that the C^* Clifford algebra $C[V]$ has a unique antiautomorphism α fixing V pointwise: for example, extend the self-adjoint Clifford map $V \to C[V]^0$ to a C^* algebra map $C[V] \overset{\alpha}{\to} C[V]^0$ by universality, where $C[V]^0$ is the C^* algebra $C[V]$ with reversed product. If $u \in \mathrm{Spin}\,_1^c(V)$ then $\alpha(u)u$ is a unitary in the centre of $C[V]$ and is therefore a unitary scalar. The resulting map

$$\nu : \mathrm{Spin}\,_1^c(V) \to \mathbb{T} : u \mapsto \alpha(u)u$$

is a unitary character of $\mathrm{Spin}\,_1^c(V)$ which we call its *spinor norm*; it satisfies

$$\mu \in \mathbb{T}, \ u \in \mathrm{Spin}\,_1^c(V) \quad \Rightarrow \quad \nu(\mu u) = \mu^2 \nu(u).$$

We may now define the C^* *spin group* $\mathrm{Spin}\,_1(V)$ to be the kernel of ν in $\mathrm{Spin}\,_1^c(V)$: thus,

$$\mathrm{Spin}\,_1(V) = \{u \in \mathrm{Spin}\,_1^c(V) : \nu(u) = 1\}.$$

The kernel of the homomorphism $\mathrm{Spin}\,_1^c(V) \to SO_1(V)$ when restricted to $\mathrm{Spin}_1(V)$ is precisely $\{\pm 1\}$ in view of the fact that ν restricts to $\mathbb{T} \subset \mathrm{Spin}_1^c(V)$ as the squaring map. Consequently, we have a central short exact sequence of groups

$$1 \to \{\pm 1\} \to \mathrm{Spin}_1(V) \to SO_1(V) \to 1$$

and the C^* spin group $\mathrm{Spin}_1(V)$ is a double cover of $SO_1(V)$.

Finally, we remark that $C[V]$ in fact contains a C^* version $\mathrm{Pin}_1(V)$ of the pin group over V and that this is made up of unitaries that are essentially real.

4.3 vN spin groups

Our chief concern in this section is to determine all inner Bogoliubov automorphisms of the vN Clifford algebra $\mathcal{A}[V]$ over an infinite-dimensional real Hilbert space V.

A simple notational device helps in both the formulation of the main result and the organization of its proof. Recall that we have been using $+$ and $-$ as parity symbols; henceforth, we shall use the *parity symbol* ε to denote either. Thus, $\mathcal{A}[V] = \mathcal{A}^+[V] \oplus \mathcal{A}^-[V]$ where $\mathcal{A}^\varepsilon[V]$ comprises all $A \in \mathcal{A}[V]$ having parity ε in the sense that $\gamma(A) = \varepsilon A$. Similarly, if J is a unitary structure then the bounded linear operator A on Fock space $\mathbb{H}_J$ is said to have parity ε if and only if $\Gamma_J A \Gamma_J = \varepsilon A$. Additionally, we shall speak of a vector space as having parity $+$ or $-$ according to whether its dimension is even or odd.

In these terms, we define the *Blattner group* $\mathcal{G} = \mathcal{G}(V)$ to be $\mathcal{G}^+ \cup \mathcal{G}^-$ where $\mathcal{G}^\varepsilon = \mathcal{G}^\varepsilon(V)$ is the set of all orthogonal transformations $g \in O(V)$ of V such that $g - \varepsilon I$ is Hilbert-Schmidt and such that $\ker(g + \varepsilon I)$ has parity ε. More fully, $\mathcal{G}^+$ comprises all $g \in O(V)$ with $g - I$ Hilbert-Schmidt and $\ker(g + I)$ even-dimensional whilst $\mathcal{G}^-$ comprises all $g \in O(V)$ with $g + I$ Hilbert-Schmidt and $\ker(g - I)$ odd-dimensional.

The main result of this section may now be formulated thus: if $g \in O(V)$ then the Bogoliubov automorphism θ_g of the vN Clifford algebra $\mathcal{A}[V]$ is inner precisely when g lies in the Blattner group $\mathcal{G}(V)$; moreover, the parity of any element of $\mathcal{G}(V)$ equals the parity of any operator in $\mathcal{A}[V]$ that implements the corresponding Bogoliubov automorphism. Of course, it follows that $\mathcal{G}(V)$ and $\mathcal{G}^+(V)$ are not only groups but also normal subgroups of $O(V)$.

Our proof begins by *doubling* the real Hilbert space V. Thus, let $V \oplus V$ be the orthogonal sum of two copies of V; of course, $V \oplus V$ is itself a real Hilbert space. In fact, $V \oplus V$ also comes equipped with a canonical unitary structure J given by

$$x, y \in V \quad \Rightarrow \quad J(x \oplus y) = (-y) \oplus x$$

so that

$$J = \begin{bmatrix} 0 & -I \\ I & 0 \end{bmatrix}$$

in block form relative to the orthogonal decomposition $V \oplus V$. We remark that the canonical map

$$V \oplus V \to V^{\mathbb{C}} : x \oplus y \to x + \mathrm{i}y$$

is a unitary isomorphism when $V \oplus V$ has the Hermitian inner product determined by J and where $V^{\mathbb{C}}$ has the Hermitian inner product extending the original inner product on V.

Recall that $\mathbb{H}_\tau$ is the complex Hilbert space completion of $H_\tau = C(V)$ in the inner product $\langle \cdot \mid \cdot \rangle_\tau$ arising from the canonical trace and that Γ is the symmetry on $\mathbb{H}_\tau$ extending the grading automorphism of the complex Clifford algebra. Recall also that λ and ρ denote respectively the left regular representation and the right regular antirepresentation of the complex Clifford algebra on either H_τ or $\mathbb{H}_\tau$. Now, define a real-linear map

$$\pi : V \oplus V \to B(\mathbb{H}_\tau)$$

by

$$\pi(x \oplus y) = \lambda(x) + \mathrm{i}\rho(y)\Gamma$$

so that

$$\pi(x \oplus y)\zeta = x\zeta + \mathrm{i}\Gamma(\zeta)y$$

for $x, y \in V$ and $\zeta \in H_\tau$. Notice that if $x, y \in V$ then

$$\begin{aligned}
\pi(x \oplus y)^2 &= \big(\lambda(x) + \mathrm{i}\rho(y)\Gamma\big)^2 \\
&= \lambda(x)^2 - \rho(y)\Gamma\rho(y)\Gamma + \mathrm{i}\rho(y)\Gamma\lambda(x) + \mathrm{i}\lambda(x)\rho(y)\Gamma \\
&= \lambda(x)^2 + \rho(y)^2 \\
&= (\|x\|^2 + \|y\|^2)I \\
&= \|x \oplus y\|^2 I
\end{aligned}$$

since $\lambda(x)$ and $\rho(y)$ anticommute with Γ and commute with each other. Notice also that if $x, y \in V$ then

$$\pi(x \oplus y)^* = \lambda(x) - \mathrm{i}\Gamma\rho(y) = \pi(x \oplus y)$$

since in addition $\lambda(x)$ and $\rho(y)$ are self-adjoint. Thus, π is a self-adjoint Clifford map and so (by the universal mapping property in Theorem 1.2.4) extends to a representation

$$\pi : C[V \oplus V] \to B(\mathbb{H}_\tau)$$

of the C^* Clifford algebra over $V \oplus V$.

Let us investigate this representation π a little more closely. The standard unit vector $\Omega := \mathbf{1}$ in $\mathbb{H}_\tau$ is certainly cyclic for π: indeed, it is already cyclic for the left regular representation λ. Moreover, if we take $w = x \oplus y \in V \oplus V$ then $\pi(w)\Omega = x + \mathrm{i}y$ and $\pi(Jw) = -y + \mathrm{i}x$ so that Ω satisfies the J-vacuum condition

$$\pi(w + \mathrm{i}Jw)\Omega = 0.$$

Thus, Ω is a cyclic unit J-vacuum vector for the representation π of $C[V \oplus V]$ on $\mathbb{H}_\tau$. It now follows from Theorem 2.4.7 that π is unitarily equivalent to the Fock representation of $C[V \oplus V]$ determined by J. The symmetry Γ plays the rôle of grading operator on $\mathbb{H}_\tau$ as Fock space, since it fixes Ω and anticommutes with $\pi(w)$ whenever $w \in V \oplus V$. Furthermore, the unitary equivalence of π with the Fock representation is rendered unique by taking Ω as Fock vacuum.

Theorem 4.3.1 *A model for the Fock representation of $C[V \oplus V]$ determined by the unitary structure*

$$J = \begin{bmatrix} 0 & -I \\ I & 0 \end{bmatrix} \in \mathbb{U}(V \oplus V)$$

is

$$\pi : C[V \oplus V] \to B(\mathbb{H}_\tau)$$

where

$$x, y \in V \quad \Rightarrow \quad \pi(x \oplus y) = \lambda(x) + \mathrm{i}\rho(y)\Gamma.$$

$\square$

The value of this result is that it makes available to us our detailed knowledge of Fock representations and the unitary implementability of Bogoliubov automorphisms within them. Before availing ourselves of this knowledge, we introduce a further piece of notation and make a further preparatory observation.

To any orthogonal transformation $g \in O(V)$ of V and to either choice of parity ε, we associate the orthogonal transformation $g_\varepsilon \in O(V \oplus V)$ of $V \oplus V$ given by

$$g_\varepsilon = \begin{bmatrix} g & 0 \\ 0 & \varepsilon I \end{bmatrix}$$

in block form relative to the orthogonal decomposition $V \oplus V$. In terms of this notation, entirely straightforward block form calculations yield the following .

Theorem 4.3.2 *If $g \in O(V)$ and if $\varepsilon \in \{+, -\}$ then*

$$[J, g_\varepsilon] = \begin{bmatrix} 0 & g - \varepsilon I \\ g - \varepsilon I & 0 \end{bmatrix}$$

and

$$g_\varepsilon - Jg_\varepsilon J = \begin{bmatrix} g + \varepsilon I & 0 \\ 0 & g + \varepsilon I \end{bmatrix}.$$

$\square$

In particular, it follows that the commutator $[J, g_\varepsilon]$ is Hilbert-Schmidt on $V \oplus V$ if and only if the difference $g - \varepsilon I$ is Hilbert-Schmidt on V, and that the (complex) parity of $\ker(g_\varepsilon - Jg_\varepsilon J)$ coincides with the (real) parity of $\ker(g + \varepsilon I)$.

Our final preparation for the proof of the main theorem is this: to observe that if $g \in O(V)$ and the automorphism θ_g of $\mathcal{A}[V]$ is inner, then any unit $U \in \mathcal{A}[V]$ implementing θ_g is either even or odd, and may be chosen to be unitary. Indeed, since θ_g and the grading automorphism γ commute, if $A \in \mathcal{A}[V]$ then

$$\begin{aligned}
\gamma(U)A\gamma(U)^{-1} &= \gamma\big(U\gamma(A)U^{-1}\big) \\
&= \gamma\big(\theta_g\gamma(A)\big) \\
&= \theta_g(A) \\
&= UAU^{-1}
\end{aligned}$$

whence $U^{-1}\gamma(U)$ lies in the centre of $\mathcal{A}[V]$. The factorial nature of $\mathcal{A}[V]$ (see Theorem 1.3.6) now implies that $\gamma(U) = \mu U$ for some $\mu \in \mathbb{C}$ and the fact that $\gamma^2 = I$ forces $\mu = \pm 1$. Thus, either, $\gamma(U) = U$ or $\gamma(U) = -U$ as claimed. It may likewise be shown that the positive operator U^*U lies in the centre of $\mathcal{A}[V]$ and is hence a scalar, which we may arrange to be $\mathbf{1}$ by rescaling. Thus U may be chosen to be unitary, as also claimed.

We may now proceed unhindered to the proof of the main result, taking its two directions separately for convenience.

Theorem 4.3.3 *Let $g \in O(V)$. If the Bogoliubov automorphism θ_g of $\mathcal{A}[V]$ is inner and implemented by a unitary in $\mathcal{A}^\varepsilon[V]$ then $g \in \mathcal{G}^\varepsilon(V)$.*

Proof Let the unitary $U \in \mathcal{A}^\varepsilon[V]$ implement θ_g and let $x, y \in V$. On the one hand, $U\lambda(x) = \lambda(gx)U$ since U implements θ_g in the left regular representation. On the other hand, $U\rho(y) = \rho(y)U$ since $U \in \mathcal{A}_\lambda = \mathcal{A}'_\rho$: see Theorem 1.3.15 or Theorem 1.3.20. These facts in hand, it follows that

$$\begin{aligned}
U\pi(x \oplus y) &= U(\lambda(x) + i\rho(y)\Gamma) \\
&= \lambda(gx)U + i\rho(y)U\Gamma \\
&= \lambda(gx)U + i\varepsilon\rho(y)\Gamma U \\
&= \pi(gx \oplus \varepsilon y)U
\end{aligned}$$

whence $U = \varepsilon\Gamma U\Gamma$ implements θ_{g_ε} in the Fock representation π of $C[V \oplus V]$ on $\mathbb{H}_\tau$. Taken together, Theorem 3.3.5 and Theorem 3.5.1 now imply that $[J, g_\varepsilon]$ is Hilbert-Schmidt and that $\ker(g_\varepsilon - Jg_\varepsilon J)$ has parity ε, whence Theorem 4.3.2 implies that $g - \varepsilon I$ is Hilbert-Schmidt and $\ker(g + \varepsilon I)$ has parity ε. We conclude by definition that $g \in \mathcal{G}^\varepsilon(V)$ as was claimed.   □

Theorem 4.3.4 *If $g \in \mathcal{G}^\varepsilon(V)$ then the Bogoliubov automorphism θ_g of $\mathcal{A}[V]$ is inner and implemented by a unitary in $\mathcal{A}^\varepsilon[V]$.*

Proof Let $g \in \mathcal{G}^\varepsilon(V)$. By definition, $g - \varepsilon I$ is Hilbert-Schmidt and $\ker(g + \varepsilon I)$ has parity ε, whence Theorem 4.3.2 tells us that $[J, g_\varepsilon]$ is Hilbert-Schmidt and $\ker(g_\varepsilon - Jg_\varepsilon J)$ has parity ε. From Theorem 3.3.5 and Theorem 3.5.1 together, we deduce the existence of a unitary operator $U = \varepsilon\Gamma U\Gamma$ on $\mathbb{H}_\tau$ such that if $x, y \in V$ then

$$U\pi(x \oplus y) = \pi(gx \oplus \varepsilon y)U.$$

On the one hand, this implies that

$$U\lambda(x) = U\pi(x \oplus 0) = \pi(gx \oplus 0)U = \lambda(gx)U$$

so that U implements θ_g in the left regular representation. On the other hand, it implies that

$$\begin{aligned}
U\rho(y) &= -\mathrm{i}U\pi(0 \oplus y)\Gamma \\
&= -\mathrm{i}\pi(0 \oplus \varepsilon y)U\Gamma \\
&= -\mathrm{i}\pi(0 \oplus y)\Gamma U \\
&= \rho(y)U
\end{aligned}$$

so that $U \in \mathcal{A}'_\rho = \mathcal{A}_\lambda = \mathcal{A}[V]$ by Theorem 1.3.20. Thus θ_g is inner, implemented by the unitary $U \in \mathcal{A}^\varepsilon[V]$.   □

Before proceeding any further, we should clear up a possible source of confusion. Recall that if $g \in O(V)$ then θ_g extends from $H_\tau = C(V)$ to define a unitary operator U_g on $\mathbb{H}_\tau$ with the property that

$$a \in C(V) \quad \Rightarrow \quad \lambda(\theta_g a) = U_g\lambda(a)U_g^*$$

and therefore

$$A \in \mathcal{A}[V] \quad \Rightarrow \quad \theta_g(A) = U_g A U_g^*;$$

see Theorem 1.3.8. We contend that the unitary operator U_g lies in the vN Clifford algebra $\mathcal{A}[V]$ if and only if $g = I$. A proof of this contention (independent of our main result) runs as follows.

Theorem 4.3.5 *If $g \in O(V)$ then the unitary operator U_g lies in $\mathcal{A}[V]$ precisely when $g = I$.*

Proof Plainly, only one direction is in question. Thus, let $g \in O(V)$ and suppose that $U_g \in \mathcal{A}[V]$. The unitary operator U_g of course fixes the standard unit vector Ω. From the fact (in Theorem 1.3.4) that Ω is separating for $\mathcal{A}[V]$ in its action on $\mathbb{H}_\tau$ it follows now that $U_g = I$ and so $g = I$ as contended. $\qquad\square$

Thus: although the Bogoliubov automorphism θ_g of $\mathcal{A}[V]$ is inner whenever g lies in the Blattner group $\mathcal{G}(V)$ and although the unitary operator $U_g \in \text{Aut } \mathbb{H}_\tau$ always implements θ_g in the left regular representation whenever g lies in the orthogonal group $O(V)$, the unitary implementer of θ_g in $\mathcal{A}[V]$ itself is never U_g unless g is the identity.

Our main result in this section, Theorem 4.3.3 and Theorem 4.3.4 combined, has direct consequences for the Blattner group. Thus: it implies that each of $O_2(V) := \mathcal{G}(V)$ and $SO_2(V) := \mathcal{G}^+(V)$ is in fact a normal subgroup of the full orthogonal group $O(V)$. For instance, let $g \in O_2(V)$ and let $U \in \mathcal{A}[V]$ be a unitary implementing $\theta_g \in \text{Aut } \mathcal{A}[V]$; if $h \in O(V)$ then $\theta_h(U)$ implements $\theta_{hgh^{-1}}$ and has the parity of U. Actually, the fact that if $h \in O(V)$ and $g \in \mathcal{G}^\pm(V)$ then $hgh^{-1} \in \mathcal{G}^\pm(V)$ follows rather easily from the definition of the Blattner group, since $hgh^{-1} - I = h(g-I)h^{-1}$ of course. It does not follow quite so easily from the definitions that $\mathcal{G}(V)$ and $\mathcal{G}^+(V)$ are subgroups of $O(V)$. However, this is evident from the main result: if g_1 and g_2 lie in $\mathcal{G}(V)$ and if U_1 and U_2 are unitaries in $\mathcal{A}[V]$ implementing θ_{g_1} and θ_{g_2} respectively, then $U_1 U_2^{-1}$ implements $\theta_{g_1 g_2^{-1}}$ and $\gamma(U_1 U_2^{-1}) = \gamma(U_1)\gamma(U_2)^{-1}$.

We are now able to introduce the version of the spin group residing within the vN Clifford algebra. Since $\mathcal{A}[V]$ is a factor, the (even) unitary elements of $\mathcal{A}[V]$ implementing the Bogoliubov automorphism θ_g induced by a given $g \in \mathcal{G}^+(V) = SO_2(V)$ are proportional by scalars of unit modulus. We write $\text{Spin}_2^c(V)$ for the group of all (even) unitaries in $\mathcal{A}[V]$ implementing Bogoliubov automorphisms induced by elements of $SO_2(V)$. A function $\text{Spin}_2^c(V) \to SO_2(V)$ is well-defined by mapping $U \in \text{Spin}_2^c(V)$ to $g \in SO_2(V)$ when θ_g is implemented by U. This assignment is quite plainly a homomorphism, yielding a central short exact sequence of groups

$$1 \to \mathbb{T} \to \text{Spin}_2^c(V) \to SO_2(V) \to 1$$

which presents $\text{Spin}_2^c(V)$ as a central extension of $SO_2(V)$ by $\mathbb{T}$. A version of the *spinor norm* again serves to pick out the real spin group itself. Let α be the unique antiautomorphism of $\mathcal{A}[V]$ that fixes V pointwise. If $U \in \text{Spin}_2^c(V)$ then $\alpha(U)U$ is a unitary in the centre of $\mathcal{A}[V]$ and is hence a unitary scalar. The resulting map

$$\nu : \mathrm{Spin}\,^c_2(V) \to \mathbb{T} : U \mapsto \alpha(U)U$$

is a unitary character whose kernel we take as the *vN spin group* $\mathrm{Spin}_2(V)$. In the light of the implication

$$\mu \in \mathbb{T}, U \in \mathrm{Spin}\,^c_2(V) \quad \Rightarrow \quad \nu(\mu U) = \mu^2 \nu(U)$$

we see that the kernel of the homomorphism $\mathrm{Spin}\,^c_2(V) \to SO_2(V)$ is precisely $\{\pm 1\}$ when restricted to $\mathrm{Spin}_2(V)$. The resulting central short exact sequence of groups

$$1 \to \{\pm 1\} \to \mathrm{Spin}_2(V) \to SO_2(V) \to 1$$

presents the vN spin group $\mathrm{Spin}_2(V)$ as a double cover of $SO_2(V)$. The vN Clifford algebra also contains a version $\mathrm{Pin}_2(V)$ of the pin group; this is a double cover of $O_2(V)$ and is comprised of essentially real operators.

The only part of the foregoing discussion of the vN spin group that requires comment is the fact that $\mathcal{A}[V]$ admits a unique antiautomorphism restricting to V as the identity; this can be seen as follows. The main conjugation κ on $C(V) = H_\tau$ extends uniquely to $\mathbb{H}_\tau$ as a conjugation operator which we continue to denote by κ for simplicity: this follows at once from the fact that if $\xi, \eta \in H_\tau$ then

$$\langle \kappa(\xi) \mid \kappa(\eta) \rangle = \tau(\overline{\eta}^* \overline{\xi})$$
$$= \overline{\tau(\eta^* \xi)}$$
$$= \overline{\langle \xi \mid \eta \rangle}.$$

We now define $\alpha : \mathcal{A}[V] \to \mathcal{A}[V]$ by

$$\alpha(A) = \kappa \circ A^* \circ \kappa$$

where $A^* \in \mathcal{A}[V]$ is the operator adjoint of $A \in \mathcal{A}[V]$ as usual. It is straightforward to check that α is an antiautomorphism of $\mathcal{A}[V]$ that extends the main antiautomorphism α of $C(V)$ identified with $\lambda\big(C(V)\big) \subset \mathcal{A}[V]$. This takes care of existence; uniqueness is obvious.

Remarks

Alternative proofs

The necessary and sufficient conditions for Bogoliubov automorphisms to be inner are approached in a variety of ways in the literature. Sufficiency in both the C^* case and the vN case is dealt with constructively in [45]: unitary implementers are constructed explicitly as infinite products (uniformly convergent in the C^* case and strongly convergent in the vN case) after a fashion reminiscent of our §1; see also [13] and [81]. Regarding necessity, we make the following few comments. In §2, we began by showing that if $g \in O(V)$ and θ_g is inner as an automorphism of $C[V]$ then either $g - I$ or $g + I$ is compact; of course, this and more follows at once from Theorem 4.3.3. The approach in [81] proceeds from the observation that inner automorphisms of $C[V]$ are universally (Fock) implemented and then applies the result quoted under "Universal implementation" in the Remarks closing Chapter Three. The fact that, if $g \in O(V)$ and θ_g is inner (for $C[V]$ or $\mathcal{A}[V]$) then at least either $g - I$ or $g + I$ is compact, may also be established by first noting that the grading automorphism γ is not inner (see Theorem 1.2.14 and Theorem 1.3.13) and then invoking a theorem of de la Harpe [44] when V is separable; indeed, this line of attack is pursued in [45].

Topological aspects

Our account has presented $\mathrm{Spin}_1(V)$ and $\mathrm{Spin}_2(V)$ simply as double covering groups over $SO_1(V)$ and $SO_2(V)$ respectively. More is true: in each case, there are natural topologies relative to which the spin group universally covers the special orthogonal group in the topological sense. In the C^* case, $SO_1(V)$ is equipped with the metric given by $d(g, h) = \|g - h\|_1$ and $\mathrm{Spin}_1(V)$ with the uniform topology. In the vN case, $SO_2(V)$ is equipped with the metric given by $d(g, h) = \|g - h\|_2$ and $\mathrm{Spin}_2(V)$ with the strong operator topology. In each case, the spin group is simply connected and the proof materially involves a consideration of the associated Lie algebras. Details may be found in [43] for the C^* case and in [63] for the vN case. Of course, these results provide satisfactory generalizations of what has long been known in the finite-dimensional situation.

Outer invariants

Recall from "vN Clifford algebras" in the Remarks for Chapter One that if V is a separable infinite-dimensional real Hilbert space then $\mathcal{A}[V]$ is a model of the hyperfinite II_1 factor. Connes [27] determined a complete set of conjugacy invariants for periodic automorphisms of the hyperfinite II_1 factor; these invariants were computed explicitly for periodic Bogoliubov automorphisms of $\mathcal{A}[V]$ by Parthasarathy & Plymen [60]. Let θ be a periodic automorphism of $\mathcal{A}[V]$ and let $p = p(\theta)$ be the least positive integer for which the automorphism θ^p is inner; if $U \in \mathcal{A}[V]$ is a unitary implementing θ^p then $\theta(U) = qU$ for some $q = q(\theta) \in \mathbb{T}$ and $q^p = 1$. The pair (p, q) is called the *outer invariant* of θ: it serves as a complete invariant for outer conjugacy; the full conjugacy invariant involves also the inner invariant, which we do not discuss. Now Parthasarathy & Plymen found that periodic Bogoliubov automorphisms of $\mathcal{A}[V]$ come with precisely the following outer invariants: either $(p, 1)$ where p is any positive integer or $(p, -1)$ where p is any even positive integer; thus, $q = \pm 1$ in any event. For details (including the inner invariant) we refer to [60].

Conventions

It is unfortunate for pin groups that if V is odd-dimensional then the Bogoliubov automorphisms of $C(V)$ induced by elements of $O(V)$ having negative determinant are not inner. An alternative convention used by some circumvents this problem, considering not the adjoint representation but rather a variant that incorporates the grading. Explicitly, the *twisted adjoint representation* $\widetilde{\mathrm{Ad}}$ of the group of units in a Clifford algebra on the algebra itself is given by the rule

$$\widetilde{\mathrm{Ad}}_u(a) = \gamma(u)au^{-1}$$

for u a unit and a in the algebra. In particular, if $v \in V$ is nonzero then $\widetilde{\mathrm{Ad}}_v$ is reflection across the hyperplane $v^{\perp}$. In finite dimensions, $O(V)$ is generated by reflections; thus, the twisted adjoint representation leads to a model for $\mathrm{Pin}(V)$ regardless of dimensional parity. The twisted adjoint representation has other attractive features: see [6]. However, it should be noted that $\widetilde{\mathrm{Ad}}$ is not an algebra automorphism unless the unit u is even: observe that $\widetilde{\mathrm{Ad}}_u(\mathbf{1}) = \gamma(u)u^{-1}$. In particular, a Bogoliubov automorphism can lie in the range of $\widetilde{\mathrm{Ad}}$ only if it is even. Accordingly, the twisted adjoint representation is not entirely suited to our purposes.

History and miscellany

In finite dimensions, the existence of a spin group covering the special orthogonal group follows from the existence of the 'exceptional' spin representation of the orthogonal Lie algebra. The Clifford algebra construction of spin groups and their representations in finite dimensions appears in the work of Brauer & Weyl [16]; rather thorough accounts over arbitrary fields are to be found in Chevalley [24] and [25]. For finite dimensions, see also [6] [23] [67] [89].

Inner Bogoliubov automorphisms of the C^* Clifford algebra were characterized in [81] by Shale & Stinespring; our account in §2 broadly follows theirs, taking ideas from Araki [3]. Inner Bogoliubov automorphisms of the vN Clifford algebra were characterized earlier in [13] by Blattner, to facilitate the representation of groups by outer automorphisms of the hyperfinite II_1 factor; our account in §3 is rather recent, being taken from [71].

Now the vN Clifford algebra $\mathcal{A}[V]$ is simply the von Neumann algebra generated by a very special quasifree representation of the C^* Clifford algebra: see under "Quasifree states" in the Remarks closing Chapter Two. It is natural to ask for a characterization of the Bogoliubov automorphisms of $C[V]$ that extend to inner automorphisms of the von Neumann algebra $\mathcal{A}_C[V]$ generated by the quasifree representation of $C[V]$ having C as its covariance. Carey [19] found that if $I + C^2$ (equivalently, $I - |C|$) is invertible, then the Bogoliubov automorphism θ_g is inner for $\mathcal{A}_C[V]$ precisely when g lies in the Blattner group $\mathcal{G}(V)$; an account compatible with our §3 is given in [73].

APPENDIX

For convenience, we set forth here the fundamentals of operator algebras: we offer only definitions and basic results without proof; details may be located quite readily in the standard texts [15] [32] [33] [50] [61] [84] [85] [87].

To begin, let A be an associative complex algebra. We say that A is unital iff it has a multiplicative identity, usually denoted by $\mathbf{1}$. The algebra A is said to be involutive (or a *-algebra) iff it is provided with an involution: that is, a conjugate-linear (or antilinear) map $A \rightarrow A$: $a \mapsto a^*$ having the properties

$$x, y \in A \quad \Rightarrow \quad (xy)^* = y^* x^*$$
$$a \in A \quad \Rightarrow \quad (a^*)^* = a;$$

a subset of A that is invariant under the involution is said to be self-adjoint. The algebra A is called a normed algebra iff it is equipped with a vector space norm $\| \cdot \|$ such that

$$x, y \in A \quad \Rightarrow \quad \|xy\| \leq \|x\| \, \|y\|$$

and is called a Banach algebra iff its norm renders it complete.

After these preparations, a C^* algebra is an involutive Banach algebra A on which the involution and norm are related by the C^* condition:

$$a \in A \quad \Rightarrow \quad \|a^* a\| = \|a\|^2.$$

The C^* algebra A is called *simple* iff 0 and A are its only norm-closed bilateral ideals and is called *central* (when unital) iff its centre comprises precisely all scalar multiples of $\mathbf{1}$; for unital C^* algebras, simple implies central.

A concrete example of a C^* algebra is furnished by $B(\mathbb{H})$: the algebra of all bounded linear operators on a complex Hilbert space $\mathbb{H}$. This is given the pointwise linear operations and has composition for product; its involution and norm are defined by stipulating that if $T \in B(\mathbb{H})$ then

$$\xi, \eta \in \mathbb{H} \quad \Rightarrow \quad \langle T\xi \mid \eta \rangle = \langle \xi \mid T^*\eta \rangle$$
$$\|T\| = \sup\{ \|T\zeta\| : \zeta \in \mathbb{H}, \ \|\zeta\| \leq 1 \}.$$

More generally, any norm-closed self-adjoint subalgebra of $B(\mathbb{H})$ is a C^* algebra. As it happens, these examples account for all C^* algebras, up to isomorphism: this is the thrust of the celebrated Gelfand-Naimark theorem, for which see below.

If A and B are involutive algebras then we call a homomorphism $\pi : A \to B$ such that $\pi(a^*) = \pi(a)^*$ for all $a \in A$ a *star-homomorphism*; if A and B are C^* algebras then we may refer to π as a C^* algebra map. It turns out that C^* algebra maps are automatically continuous: indeed, if $\pi : A \to B$ is a C^* algebra map then $\|\pi(a)\| \leq \|a\|$ for all $a \in A$; further, if π is injective then it is isometric in the sense that $\|\pi(a)\| = \|a\|$ for all $a \in A$. In these terms, the C^* algebra A is simple iff each C^* algebra map defined on A is isometric.

We call a star-homomorphism $\pi : A \to B(\mathbb{H})$ a star-representation (or just a representation) of the involutive algebra A on the complex Hilbert space $\mathbb{H}$; as usual, to say that the representation π is faithful means that the homomorphism π is injective. Now, the Gelfand-Naimark theorem (to which we alluded above) asserts that each C^* algebra A has a faithful representation $\pi : A \to B(\mathbb{H})$ on some complex Hilbert space; the range of π is a norm-closed self-adjoint subalgebra of $B(\mathbb{H})$ to which A is isomorphic via π.

Let A be an involutive algebra of which $\pi : A \to B(\mathbb{H})$ is a star-representation. We say that π is *irreducible* iff it satisfies either (hence both) of the following equivalent conditions: that the only closed subspaces of $\mathbb{H}$ stable under $\pi(a)$ for each $a \in A$ are 0 and $\mathbb{H}$ itself; that the only elements of $B(\mathbb{H})$ commuting with $\pi(a)$ for each $a \in A$ are the scalar operators. We call π *cyclic* iff there is a vector $\zeta \in \mathbb{H}$ that is cyclic for π in the sense that the subspace $\{\pi(a)\zeta : a \in A\}$ is dense in $\mathbb{H}$. These notions are connected by the fact that π is irreducible iff each nonzero vector in $\mathbb{H}$ is cyclic.

A state on the (for convenience) unital C^* algebra A is a linear functional $\phi : A \to \mathbb{C}$ that is positive in the sense that $\phi(a^*a) \geq 0$ whenever $a \in A$ and normalized by $\phi(1) = 1$. If $\pi : A \to B(\mathbb{H})$ is a star-representation and $\zeta \in \mathbb{H}$ a unit vector, then

$$\phi : A \to \mathbb{C} : a \mapsto \langle \pi(a)\zeta \mid \zeta \rangle$$

is a state on A. In the opposite direction, via the Gelfand-Naimark-Segal (GNS) construction, to each state $\phi : A \to \mathbb{C}$ there is naturally associated a star-representation $\pi : A \to B(\mathbb{H})$ with a cyclic unit vector $\zeta \in \mathbb{H}$ such that $\langle \pi(a)\zeta \mid \zeta \rangle = \phi(a)$ whenever $a \in A$. Let the state ϕ correspond to the cyclic representation π in this way; then π is irreducible iff ϕ is pure in the sense that it is an extreme point in the convex set of all states on A.

This concludes what we have to say about C^* algebras in general. We next make some remarks concerning a special class of C^* algebras concretely represented as algebras of bounded linear operators on complex Hilbert spaces, namely, von Neumann (or vN) algebras.

Let $\mathbb{H}$ be a complex Hilbert space. The commutant of the subset $S \subset B(\mathbb{H})$ is the collection $S' \subset B(\mathbb{H})$ comprising all bounded linear operators on $\mathbb{H}$ that commute with each element of S; the bicommutant of S is its double commutant $S'' = (S')'$. We define a von Neumann algebra on $\mathbb{H}$ to be a self-adjoint subalgebra A of $B(\mathbb{H})$ such that $A'' = A$; with this definition, a von Neumann algebra contains the identity operator and is hence automatically unital. The von Neumann algebra A is called a factor iff it is central in that its centre consists precisely of the scalar operators: $A \cap A' = \mathbb{C}\mathbf{1}$.

A von Neumann algebra is certainly norm-closed and hence in particular a C^* algebra. In fact, a von Neumann algebra on $\mathbb{H}$ is closed in a number of other standard locally convex topologies on $B(\mathbb{H})$; here, we mention only two of these operator topologies. The weak operator topology on $B(\mathbb{H})$ is that defined by the seminorms

$$B(\mathbb{H}) \to \mathbb{R} : T \mapsto | \langle T\xi \mid \eta \rangle |$$

as ξ and η run over $\mathbb{H}$: the net $(T_j : j \in \mathcal{J})$ in $B(\mathbb{H})$ converges to the operator $T \in B(\mathbb{H})$ in the weak operator topology (written $T_j \overset{w}{\to} T$) iff $\langle T_j\xi \mid \eta \rangle \to \langle T\xi \mid \eta \rangle$ for all ξ and η in $\mathbb{H}$. The ultraweak (or σ-weak) operator topology on $B(\mathbb{H})$ is determined by the seminorms

$$B(\mathbb{H}) \to \mathbb{R} : T \mapsto \sum_{n>0} | \langle T\xi_n \mid \eta_n \rangle |$$

for sequences $(\xi_n : n \in \mathbb{N})$ and $(\eta_n : n \in \mathbb{N})$ in $\mathbb{H}$ with

$$\sum_{n>0} \left(\|T\xi_n\|^2 + \|T\eta_n\|^2 \right) < \infty.$$

Now, a von Neumann algebra $A = A''$ on $\mathbb{H}$ is closed in both the weak and the ultraweak topologies; conversely, a self-adjoint subalgebra $A \ni \mathbf{1}$ of $B(\mathbb{H})$ closed in either the weak or the ultraweak topology is a von Neumann algebra. This fact is a part of the important von Neumann bicommutant theorem, the full version of which refers not only to the weak and ultraweak topologies but also to several others.

Let A and B be von Neumann algebras and $\pi : A \to B$ a star-homomorphism. Of course, π is norm-continuous because von Neumann algebras are in particular C^* algebras. It might or might not be the case that π is continuous for the weak operator topologies on A and B. When π is weakly continuous, it is necessarily normal in the sense that if $(a_j : j \in \mathcal{J})$ is a bounded increasing net of self-adjoint elements with supremum a in A then $(\pi(a_j) : j \in \mathcal{J})$ has supremum $\pi(a)$ in B. When π is normal, it is necessarily continuous for the ultraweak topologies on A and B. Isomorphisms between von Neumann algebras are always normal and hence ultraweakly continuous. We remark that if $\pi : A \to B(\mathbb{H})$ is a normal star-representation of the von Neumann algebra A for which $\pi(\mathbf{1}) = \mathbf{1}$ then its range $\pi(A)$ is a von Neumann algebra on the complex Hilbert space $\mathbb{H}$. We remark also that the notion of normality has a counterpart for states; this counterpart is equivalent to ultraweak continuity.

References

[1] H. Araki: On the diagonalization of a bilinear Hamiltonian by a Bogoliubov transformation, *Publ. RIMS Kyoto Univ.* **4** (1968) 387–412

[2] H. Araki: On quasifree states of CAR and Bogoliubov automorphisms, *Publ. RIMS Kyoto Univ.* **6** (1970/71) 385–442

[3] H. Araki: Bogoliubov automorphisms and Fock representations of canonical anticommutation relations, *Amer. Math. Soc. Contemporary Mathematics* **62** (1987) 23–141

[4] H. Araki & D.E. Evans: On a C^* algebra approach to phase transition in the two-dimensional Ising model, *Comm. Math. Phys.* **91** (1983) 489–503

[5] E. Artin: *Geometric algebra*, Wiley-Interscience (1957)

[6] M.F. Atiyah, R. Bott & A. Shapiro: Clifford modules, *Supplement, Topology* **3** (1964) 3–38

[7] J.C. Baez, I.E. Segal & Z. Zhou: *Introduction to algebraic and constructive quantum field theory*, Princeton (1992)

[8] B.M. Baker: Free states of the gauge invariant canonical anticommutation relations, *Trans. Amer. Math. Soc.* **237** (1978) 35–61

[9] E. Balslev, J. Manuceau & A. Verbeure: Representations of anticommutation relations and Bogoliubov transformations, *Comm. Math. Phys.* **8** (1968) 315–326

[10] E. Balslev & A. Verbeure: States on Clifford algebras, *Comm. Math. Phys.* **7** (1968) 55–76

[11] C. Barnett, R.F. Streater & I. Wilde: The Itô-Clifford integral, *J. Functional Analysis* **48** (1982) 172–212

[12] F.A. Berezin: *The method of second quantization*, Academic Press (1966)

[13] R.J. Blattner: Automorphic group representations, *Pacific J. Math.* **8** (1958) 665–677

[14] N. Bourbaki: *Topological Vector Spaces, Chapters 1–5*, Spinger-Verlag (1987)

[15] O. Bratteli & D.W. Robinson:*Operator algebras and quantum statistical mechanics*, Springer-Verlag, *Volume I* (1979), *Volume II* (1981)

[16] R. Brauer & H. Weyl: Spinors in n dimensions, *Amer. J. Math.* **57** (1935) 425–449

[17] A.L. Carey: Some infinite dimensional groups and bundles, *Publ. RIMS Kyoto Univ.* **20** (1984) 1103–1117

[18] A.L. Carey: Projective representations of the Hilbert Lie group $\mathcal{U}(H)_2$ via quasifree states on the CAR algebra, *J. Functional Analysis* **55** (1984) 277–296

[19] A.L. Carey: Inner automorphisms of hyperfinite factors and Bogoliubov transformations, *Ann. Inst. Henri Poincaré, Phys. théor.* **40** (1984) 141–149

[20] A.L. Carey, C.A. Hurst & D.M. O'Brien: Automorphisms of the canonical anticommutation relations and index theory, *J. Functional Analysis* **48** (1982) 360–393

[21] A.L. Carey & D.M. O'Brien: Automorphisms of the infinite-dimensional Clifford algebra and the Atiyah-Singer mod 2 index, *Topology* **22** (1983) 437–448

[22] A. Carey & J. Palmer: Infinite complex spin groups, *J. Functional Analysis* **83** (1989) 1–43

[23] E. Cartan: *The theory of spinors*, Dover (1981)

[24] C. Chevalley: *The algebraic theory of spinors*, Columbia University Press (1954)

[25] C. Chevalley: *The construction and study of certain important algebras*, Publ. Math. Soc. Japan **1** (1955)

[26] W.K. Clifford: Applications of Grassmann's extensive algebra , *Amer. J. Math.* **1** (1878) 350–358

[27] A. Connes: Periodic automorphisms of the hyperfinite factor of type II_1, *Acta. Sci. Math.* **39** (1977) 39–66

[28] J. Cook: The mathematics of second quantization, *Trans. Amer. Math. Soc.* **74** (1953) 222–245

[29] G.F. Dell'Antonio: Structure of the algebras of some free systems, *Comm. Math. Phys.* **9** (1968) 81–117

[30] J. Dieudonné: *Histoire de mathématiques, 1700–1900*, Hermann, Paris (1978)

[31] P.A.M. Dirac: The quantum theory of the electron, *Proc. Roy. Soc. A* **117** (1928) 610–629

[32] J. Dixmier: *Von Neumann algebras*, Elsevier North-Holland (1981)

[33] J. Dixmier: C^*-*algebras*, Elsevier North-Holland (1982)

[34] S. Doplicher & R.T. Powers: On the simplicity of the even CAR algebra and free field models, *Comm. Math. Phys.* **7** (1968) 77–79

[35] G.G. Emch: *Algebraic methods in statistical mechanics and quantum field theory*, Wiley-Interscience (1972)

[36] V. Fock: Konfigurationsraum und zweite quantelung, *Z. Phys.* **75** (1932) 622–647

[37] K. Fredenhagen: Implementation of automorphisms and derivations of the CAR algebra, *Comm. Math. Phys.* **52** (1977) 255–266

[38] K.O. Friedrichs: *Mathematical aspects of the quantum theory of fields*, Interscience (1953)

[39] L. Gårding & A.S. Wightman: Representations of the commutation and anticommutation relations, *Proc. Nat. Acad. Sci.* **40** (1954) 617–626

[40] J.F. Gille: On a question of André Verbeure, *Comm. Math. Phys.* **34** (1973) 131–134

[41] J. Glimm & A. Jaffe: *Quantum physics*, Second edition, Springer-Verlag (1987)

[42] A. Guichardet: Produits tensoriels infinis et représentations des relations d'anticommutation, *Ann. Sci. Ec. Norm. Sup.* **83** (1966) 1–52

[43] P. de la Harpe: The Clifford algebra and the spinor group of a Hilbert space, *Compositio Math.* **25** (1972) 245–261

[44] P. de la Harpe: Sous-groupes distingués du group unitaire et du groupe général linéaire d'un espace de Hilbert, *Comment. Math. Helv.* **51** (1976) 241–257

[45] P. de la Harpe & R.J. Plymen: Automorphic group representations: a new proof of Blattner's theorem, *J. London Math. Soc.* **19** (1979) 509–522

[46] R.H. Herman & M. Takesaki: States and automorphism groups of operator algebras, *Comm. Math. Phys.* **19** (1970) 142–160

[47] N.M. Hugenholtz & R.V. Kadison: Automorphisms and quasi-free states on the CAR algebra, *Comm. Math. Phys.* **43** (1975) 181–197

[48] A. Jaffe, A. Lesniewski & J. Weitsman: Pfaffians on Hilbert space, *J. Functional Analysis* **83** (1989) 348–363

[49] P. Jordan & E. Wigner: Über das Paulische Äquivalenzverbot, *Z. Phys.* **47** (1928) 631–651

[50] R.V. Kadison & J.R. Ringrose: *Fundamentals of the theory of operator algebras*, Academic Press, Volume I (1983), Volume II (1986)

[51] B. Kostant & S. Sternberg: Symplectic reduction, BRS cohomology, and infinite dimensional Clifford algebras, *Ann. Phys.* **176** (1987) 40–113

[52] G. Labonté: On the nature of "strong" Bogoliubov transformations for fermions, *Comm. Math. Phys.* **36** (1974) 59–72

[53] S. Lang: *Algebra*, Addison-Wesley (1965)

[54] L.-E. Lundberg: Quasi-free "second quantization", *Comm. Math. Phys.* **50** (1976) 103–112

[55] J. Manuceau, F. Rocca & D. Testard: On the product form of quasi-free states, *Comm. Math. Phys.* **12** (1969) 43–57

[56] J. Manuceau & A. Verbeure: Non-factor quasi-free states of the CAR algebra, *Comm. Math. Phys.* **18** (1970) 319–326

[57] J. Manuceau & A. Verbeure: The theorem on unitary equivalence of Fock representations, *Ann. Inst. Henri Poincaré (A)* **16** (1971) 87–91

[58] T. Matsui: On quasi-equivalence of quasi-free states of gauge-invariant CAR algebras, *J. Operator Theory* **17** (1987) 281–290

[59] J. von Neumann: Continuous geometry, *Proc. Nat. Acad. Sci.* **22** (1936) 92–100

[60] K.R. Parthasarathy & R.J. Plymen: On the outer and inner invariants of Connes, *J. Functional Analysis* **38** (1980) 1–15

[61] G.K. Pedersen: *C^* algebras and their automorphism groups*, Academic Press (1979)

[62] R.J. Plymen: Spinors in Hilbert space, *Math. Proc. Camb. Phil. Soc.* **80** (1976) 337–347

[63] R.J. Plymen & R.F. Streater: A model of the universal covering group of $SO(E)_2$, *Bull. London Math. Soc.* **7** (1975) 283–288

[64] R.J. Plymen & R.M.G. Young: On the spin algebra of a real Hilbert space, *J. London Math. Soc.* **9** (1974) 286–292

[65] R.T. Powers: Representations of the canonical anticommutation relations, Princeton thesis (1967)

[66] R.T. Powers & E. Størmer: Free states of the canonical anticommutation relations, *Comm. Math. Phys.* **16** (1970) 1–33

[67] A. Pressley & G. Segal: *Loop groups*, Oxford University Press (1986)

[68] C.R. Putnam & A. Wintner: The orthogonal group in Hilbert space, *Amer. J. Math.* **74** (1952) 52–78

[69] G. Rideau: On some representations of the anticommutation relations, *Comm. Math. Phys.* **9** (1968) 229–241

[70] P.L. Robinson: The even C^* Clifford algebra, *Proc. Amer. Math. Soc.* **118** (1993) 713–714

[71] P.L. Robinson: Yet another proof of Blattner's theorem, *John Klauder anniversary volume*, World Scientific (to appear)

[72] P.L. Robinson: Quasifree representations of Clifford algebras, *Math. Proc. Camb. Phil. Soc.* **113** (1993) 487–497

[73] P.L. Robinson: Modular theory and Bogoliubov automorphisms of Clifford algebras, *J. London Math. Soc.* **49** (1994) 463–476

[74] F. Rocca, M. Sirugue & D. Testard: On a class of equilibrium states under the Kubo-Martin-Schwinger boundary condition. I. Fermions, *Comm. Math. Phys.* **13** (1969) 317–334

[75] S.N.M. Ruijsenaars: On Bogoliubov transformations for systems of relativistic charged particles, *J. Math. Phys.* **18** (1977) 517–526

[76] S.N.M. Ruijsenaars: On Bogoliubov transformations. II. The general case, *Ann. Phys.* **116** (1978) 105–134

[77] R. Schrader & D.A. Uhlenbrock: Markov structures on Clifford algebras, *J. Functional Analysis* **18** (1975) 369–413

[78] I.E. Segal: Tensor algebras over Hilbert spaces II, *Ann. Math.* **63** (1956) 160–175

[79] I.E. Segal: *Mathematical problems of relativistic physics*, Amer. Math. Soc. (1963)

[80] D. Shale & W.F. Stinespring: States of the Clifford algebra, *Ann. Math.* **80** (1964) 365–381

[81] D. Shale & W.F. Stinespring: Spinor representations of infinite orthogonal groups, *J. Math. Mech.* **14** (1965) 315–322

[82] E. Størmer: The even CAR-algebra, *Comm. Math. Phys.* **16** (1970) 136–137

[83] R.F. Streater & A.S. Wightman: *PCT, spin and statistics, and all that*, Addison-Wesley Advanced Book Classics (1989)

[84] V.S. Sunder: *An invitation to von Neumann algebras*, Springer-Verlag (1987)

[85] M. Takesaki: *Theory of operator algebras I*, Springer-Verlag (1979)

[86] B. Thaller: *The Dirac equation*, Springer-Verlag (1992)

[87] D.M. Topping: *Lectures on von Neumann algebras*, Van Nostrand Reinhold (1971)

[88] B.L. van der Waerden: *A history of algebra*, Springer-Verlag (1985)

[89] G.W. Whitehead: *Elements of homotopy theory*, Springer-Verlag (1978)

[90] J.C. Wolfe: Free states and automorphisms of the Clifford algebra, *Comm. Math. Phys.* **45** (1975) 53–58

Index